Biological Electron Microscopy

Biological Electron Microscopy

Barbra L. Gabriel

VNR VAN NOSTRAND REINHOLD COMPANY
NEW YORK CINCINNATI TORONTO LONDON MELBOURNE

Library of Congress Catalog Card Number: 82-6946
ISBN: 0-442-22923-2

Manufactured in the United States of America

Published by Van Nostrand Reinhold Company Inc.
135 West 50th Street, New York, N.Y. 10020

Van Nostrand Reinhold Publishing
1410 Birchmount Road
Scarborough, Ontario M1P 2E7, Canada

Van Nostrand Reinhold Australia Pty. Ltd.
17 Queen Street
Mitcham, Victoria 3132, Australia

Van Nostrand Reinhold Company Limited
Molly Millars Lane
Wokingham, Berkshire, England

15 14 13 12 11 10 9 8 7 6 5 4 3 2 1

Library of Congress Cataloging in Publication Data

Gabriel, Barbra L.
Biological electron microscopy.

Includes index.
1. Electron microscopic, Transmission. 2. Electron microscopy. I. Title. [DNLM: 1. Microscopy, Electron. QH 212.E4 G118b]
QH212.T7G3 578′.45 82-6946
ISBN 0-442-22923-2 AACR2

To AMG and ARG

Preface

The transmission electron microscope is an integral educational and research tool for the life scientist. Modern undergraduate and graduate biological curricula routinely include reference to, if not actual use of, the TEM. Pathologists have shown that the TEM is an invaluable tool for the exact diagnosis of various diseases, and researchers continue to explore biological ultrastructure. In light of the need to properly educate potential microscopists, as well as provide a knowledgeable rather than cookbook approach to microscopy, this introductory level text has been assembled. Hopefully it will serve the purpose of combining the theoretical comprehension necessary for effective application of various techniques.

The author has attempted to give full credit to the researchers who provided the entire framework for this text; for those who were inadvertently not cited, I sincerely apologize for the omission.

BARBRA GABRIEL

Acknowledgments

My gratitude extends to my colleagues who shared their knowledge, offered encouragement, and judicially critiqued my work. Special thanks to my reviewers, Dr. Fredric Giere and Dr. Bruce Murray, and to the editors of Van Nostrand Reinhold for their guidance.

Contents

1. Transmission Electron Microscope Instrumentation

INTRODUCTION

The electron microscope (EM) is a powerful tool for the study of ultrastructure because it combines high resolution with high magnification. Light microscopy readily resolves individual cells and their nuclei; electron microscopy probes deeper into the cell, allowing examination of organelles such as mitochondria which are only hazy dots in the light microscope. The magnitude of order defining each microscope is a function of the type of irradiation—light or electrons—used to form an image.

The similarity between light and electron microscopes is clear. In both, the source of radiation is usually a tungsten filament. By definition, a light microscope uses light for imaging, and an electron microscope, electrons. Glass or quartz lenses focus light, and electromagnetic lenses focus electrons onto the specimen; the resulting interactions give rise to an image. Observation of a light image may be made directly by the eye, whereas electrons are converted into a light image by striking a fluorescent screen.

The basic instrument considered here is transmission; that is, the light or electron signal is detected after it has passed through the specimen. This is the opposite of reflection, as in reflected light microscopy and scanning electron microscopy; in these methods the surface of opaque specimens is observed.

The transmission electron microscope (TEM) consists of four systems:

1. The illuminating system, which serves to produce the required radiation and direct it onto the specimen.
2. The imaging system, composed of a series of lenses that produce the final magnified image of the specimen.
3. The image recording system, which converts the electron image observed on a fluorescent screen into a permanent record, or electron micrograph.
4. The high vacuum system, to reduce the amount of gases within the microscope which would otherwise degrade resolution by interacting with the electron beam.

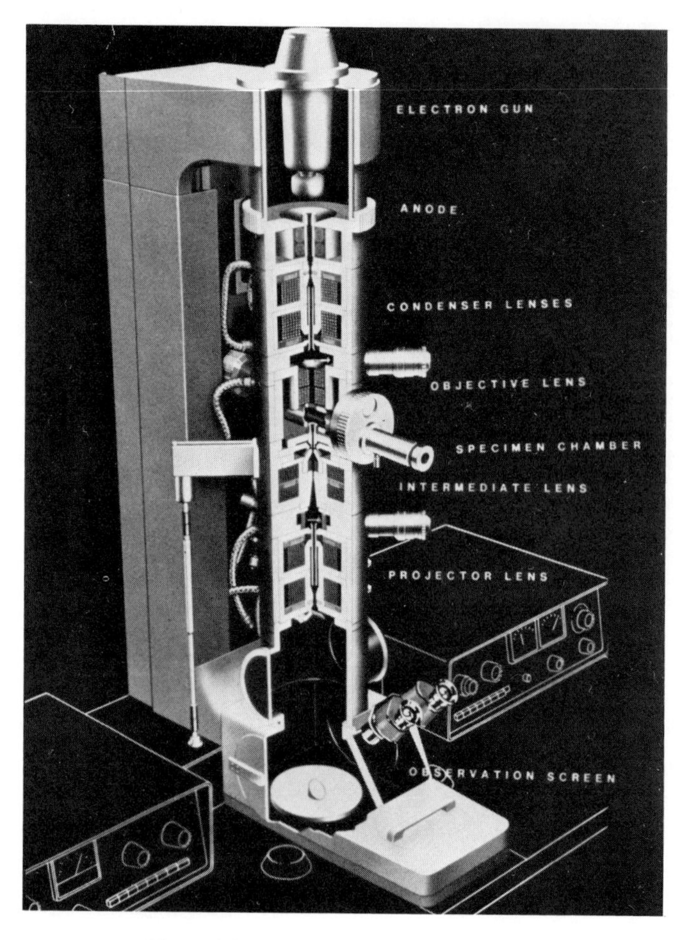

Figure 1-1. Cutaway view of Zeiss EM 10.

Each of these sytems will be discussed below, as will the factors controlling resolution. Reference to Figure 1-1 will help define these systems.

ILLUMINATING SYSTEM

The illuminating system serves to produce the electron beam and direct it onto the sample. It consists of (1) the electron gun which generates electrons, and (2) the condenser lens assembly which directs the electron beam onto the sample (Figure 1-2). The electron gun may be further broken down into the filament, shield, and anode.

The major prerequisite for the filament is that it be composed of a material capable of generating an intense beam of electrons. Most microscopes meet these requirements by using a V-shaped tungsten

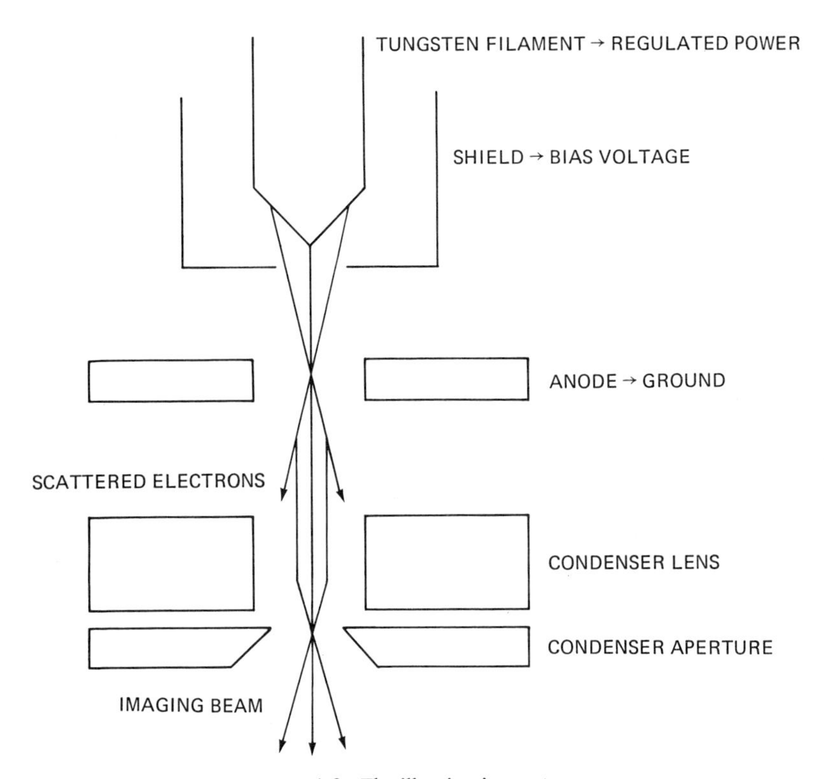

Figure 1-2. The illuminating system.

wire, about 0.2 mm in diameter, which can be electrically heated to incandescence. The filament is held at approximately 20,000 V below ground potential, and is thus the equivalent of a cathode. At low heater currents, electrons are drawn from favorable crystalline positions within the filament, forming a multiple electron source. This condition is referred to as undersaturation. As the temperature of the filament is increased (Figure 1-3), the emitting areas correspondingly increase until the filament's tip symmetrically generates electrons. This is the condition of saturation, or the effective electron source; further heating of the filament has no effect on intensity but will drastically reduce the lifetime of the filament (Figure 1-3).

Surrounding the filament and still part of the electron gun is an apertured cylinder known as the grid cap (synonyms: shield and Wehneldt cylinder). The aperture is 1 to 3 mm in diameter and must be centered over the filament tip. The distance separating the tip from the aperture greatly influences beam brightness because a small potential difference, or bias, is applied across that gap. Thus electrons will be attracted from the filament toward the grid cap, and accelerated from this point by an even greater potential difference

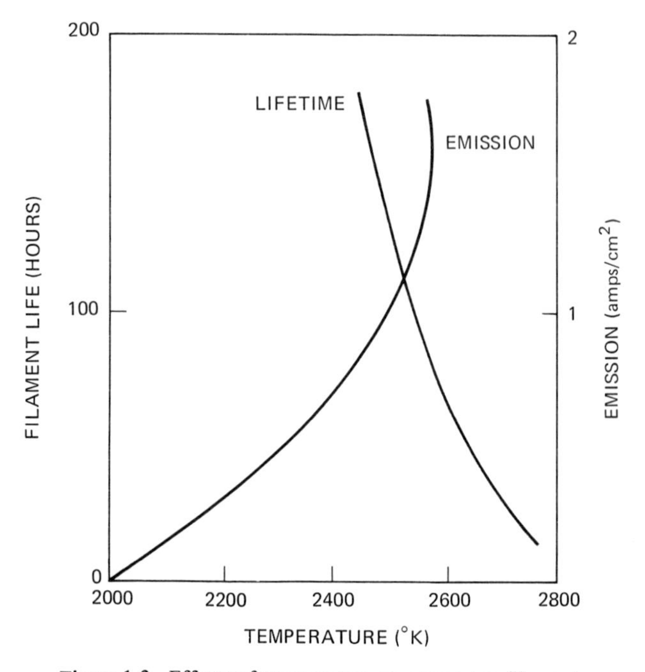

Figure 1-3. Effects of temperature on a tungsten filament.

toward the anode. Hence the anode is at ground potential but at a very large negative potential with respect to the filament; coaxial alignment of the filament/grid cap/anode gives rise to an accelerated beam of electrons. After leaving the electron gun, the electrons travel down the microscope column at constant velocity.

Routine maintenance of the electron gun should occur whenever a filament burns out (i.e., no current registers when the current is increased). Filament lifetime varies widely, from 20 to 300 hr depending upon the typical operating vacuum and general operator attention. For example, one must allow about 5 min. cool-down time before coming to atmosphere (e.g., when changing specimens). Otherwise, gas will oxidize and erode the hot filament and severely limit its lifetime to only a few hours. Vacuum leaks, contamination, and overheating destroy filaments, and all these problems are within operator control.

Specifications for changing a filament are included with the manufacturer's operating manual for the microscope, but a few general rules apply.

1. Always wear lint-free cotton or nylon gloves when handling the gun assembly. Also necessary are cotton-tipped applicators, lint-free cleaning cloth, acetone, compressed Freon or air, and metal polish.
2. The inside of the grid cap usually shows a bluish discoloration from the evaporation of tungsten and/or contamination. This is removed as follows:
   a. Using a commercially available metal polish amenable to vacuum exposure (e.g., Pol Metalputz) and a lint-free cloth, polish the grid cap until it is rid of all discoloration. Clean and polish the grid cap aperture with cotton applicators. If any discoloration exists on the outside of the shield, remove it.
   b. Remove excess polish with acetone, giving particular attention to the aperture.
   c. If possible, place the grid cap in a small beaker of acetone and then in an ultrasonic bath. Rinse it in clean acetone.
   d. Handle the grid cap with gloves.
3. Observe the used filament with a binocular light microscope.
   a. A rough, eroded filament is normal, even after a reasonable

lifetime. If the filament burned out within 15 hr of use, check for vacuum leaks, or increase cool-down time.
   b. If the broken ends of the filament are beaded, tungsten is melting, indicating a high voltage problem.
   c. If no current registers on the indicator meter and the filament shows no damage, clean the filament electrodes (trace contaminants may prevent contact).
4. Clean all exposed parts of the electron gun with compressed gas or Freon. If there is excess contamination, especially oil in older instruments, wipe it with acetone.
5. Never leave a gun assembly open to the atmosphere because severe contamination is inevitable. Do not stop anywhere in this procedure; continue through as one operation.

The second component of the illuminating system is the condenser lens assembly. It serves the dual function of demagnifying the electron beam and focusing it onto the sample. The diameter of the beam as it exits the electron gun is 25,000 to 50,000 Å, but after passing through one condenser lens it is reduced to about 100 Å. This lens is equipped with a variable aperture that intercepts stray electrons, thus preventing background scattering and reducing noise in the signal. Consequently, the effective electron gun current of about 10^{-4} amp (roughly equal to 10^{15} electrons/sec) is reduced to 10^{-12} to 10^{-10} amp (or $.6 \times 10^{-6}$ electrons/sec) at the specimen level.

All modern TEMs have a double condenser lens system. The first lens form a highly demagnified image of the electron source, and the second projects this image into the plane of the objective lens. Older instruments having only a single condenser combine both functions.

The effect of varying the focal length of the condenser is to vary brightness. Its brightest point—usually the middle of the range—is referred to as crossover. This position is important because it is used during electron-optical alignment of the microscope column, and may be used to define filament saturation. These operations, readily observed in the microscope, are described as follows:

1. The meter will show an increase in the current as the filament is heated gradually, and there will be a faint illumination of the screen (Figure 1-4a).

2. A further increase in the current will simultaneously increase the image brightness.
3. Adjusting the beam brightness (focal length of the condenser lens) to crossover will reveal a very bright image.
4. Altering the filament current will yield a bright, striped image surrounded by a hazy but symmetrical halo (Figure 1-4b and 1-4c). This point is undersaturation. (The striped image is the tip of the filament.)

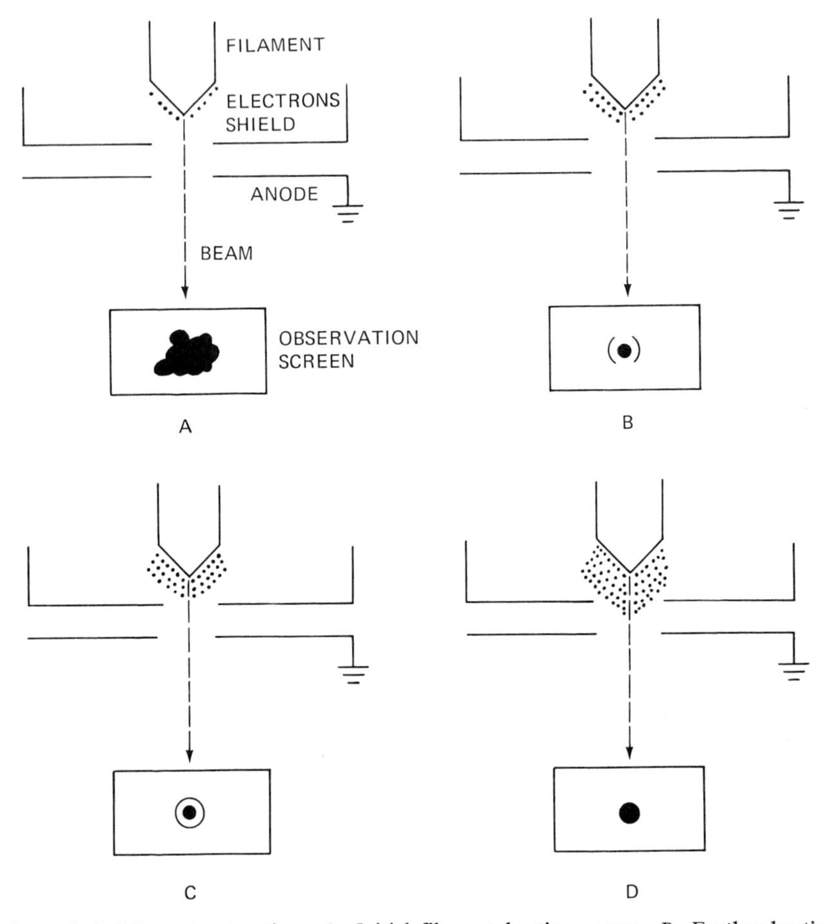

Figure 1-4. Filament saturation. A. Initial filament heating occurs. B. Further heating and brightness adjustment demonstrate crossover. C. Increasing current gives rise to a bright spot symmetrically surrounded by a halo: undersaturation. D. At saturation, the entire filament tip contributes to forming a homogeneous beam.

5. Then increasing the filament current will cause these separate images to coalesce into a very bright symmetrical spot (Figure 1-4d). Adjust the beam brightness control, and observe that this homogeneous image has various degrees of illumination and that the tip of the filament is no longer visible. This point is saturation; no further increase of filament current will increase brightness. Note the meter reading of the current for future reference.
6. There is misalignment of the illuminating system if the halo is asymmetrical during undersaturation. To obtain symmetry, move the gun alignment controls in the direction of the brighter halo, and return to center with the condenser alignment controls. Repeat until symmetry is established.

IMAGING SYSTEM

The second system of the TEM consists of a series of electromagnetic lenses that produce the final magnified image of the specimen. These are the objective, intermediate, and projector lenses.

The objective lens forms the initial enlarged image of the illuminated portion of the specimen in a plane that is suitable for further magnification by the intermediate lens. Because it sets the limit of resolution of the microscope, the objective lens is its most critical component. Image focusing involves changing the focal length of the objective lens (i.e., altering the current that passes through the magnetic coil). Since the adjustments in focal length may vary over a wide range, several controls of progressively increasing sensitivity are provided to afford the precision in focusing that is required.

All TEMs are equipped with an optical microscope (10X magnification) at the observation level, which must be used to fine-focus an image for photography. The optical microscope is focused on an "x" penciled or scratched on the fluorescent screen. This microscope can then be used to assess the focus of the electron image falling on that same plane. Ensuring that the electron image is focused at this point permits enlargement of the negative during printing.

The specimen holder is placed in the field of the objective lens. Most microscopes are provided with a number of sample holders of varying size; short holders are used for low magnification work (about 1000X), a medium-size holder is used for intermediate ranges, and a long holder is used for very high magnification work (greater than 250,000X). These various sizes permit extension of the focal length of the objective lens, allowing a high degree of freedom of magnification.

The objective lens is equipped with a molybdenum or platinum strip-type aperture that has pores of various diameters. Like the condenser lens aperture, this device intercepts excess electrons that have been scattered by the sample. It greatly influences the contrast of the final image: as a rule, the smaller the pore size the higher the contrast. Contrast will be discussed later, but ideally one wants a range of black/gray tones in an electron micrograph.

Both the condenser and objective apertures must be cleaned routinely, as contamination tends to accumulate at the perimeter and degrade resolution (by inducing spherical aberration). Molybdenum apertures are cleaned in a high vacuum bell-jar system by heating them to just below the melting point; platinum apertures are chemically cleaned by boiling them in concentrated sulfuric acid. The sample holder is also subject to a great deal of contamination by handling, repeated exposure to the atmosphere, and specimen sublimation. Clean the sample holder routinely with acetone, and if severe discoloration is present, with metal polish. Always handle the specimen holder with gloves or a cloth.

Some microscopes are equipped with an anti-contamination device, or cold trap, in the area near the sample chamber. This is a liquid-nitrogen-cooled metal rod referred to as a cold-finger. Contamination will be attracted to this point, thus preventing its accumulation at the specimen level.

Immediately beneath the objective lens is a series of additional lenses that further enlarge the image. The first is the intermediate lens, which magnifies a portion of the image emerging from the objective lens and directs it toward the projector lens. Most microscopes have only one intermediate lens; those others that are capable of very high magnification (e.g., 800,000X) may have two or three intermediate lenses.

Very similar to this lens is the projector lens, which further enlarges a portion of the intermediate image and projects it onto the observation screen. It has a much longer focal length than the other imaging lenses, meaning that an image focused on the observation screen will remain in focus up to a meter beneath the screen. Thus, the image recorded in the film chamber and that observed at the fluorescent screen are both in focus.

Magnification (M) in the light microscope is the product of the magnifications of the objective lens and the ocular; similarly, in the TEM final magnification is the product of the objective, intermediate, and projector lenses:

$$M_{\text{image}} = M_{OL} \times M_{IL} \times M_{PL}$$

Another important relationship expresses the magnification required for a given resolution:

$$M = \frac{de}{d}$$

where

de = resolution of the unaided eye
= 0.2 mm
d = resolution of the TEM

For example, for the unaided eye to resolve 10 Å in the microscope, a minimum magnification of 200,000X is required; that is, at lower magnifications human eyes cannot resolve better than 10 Å. Resolution improves significantly when an enlarged electron micrograph is viewed because film is able to record finer data than those observed in the TEM (to be discussed).

The manufacturer of a TEM normally specifies instrument magnifications within an error range of ±10%. In some situations (e.g., sizing of particles) it may be necessary to determine the exact microscope magnification. A variety of standards are commercially available for this, the most common being latex spheres for low magnification, grating replicas for intermediate magnification, and catalase crystals for high magnification. Regardless of the standard used, at each magnification setting, several micrographs are recorded to lessen statistical errors. The following calculation using a grating replica defines magnification as:

$$M = \frac{xg}{25{,}400y \times 10^3}$$

where

x = total distance between limiting lines (mm)

g = no. lines/inch of grating (54,864)
y = no. spaces counted between limiting lines

IMAGE RECORDING SYSTEM

The observation screen and photographic film comprise the image recording system. Because electrons are invisible to the human eye, the screen and film essentially translate that signal into light. Fluorescent screens coated with a 50–100-μm layer of zinc and cadmium sulfides are used for localization, focusing, and observation of the image. The resolution of the image at this level depends primarily upon the native granularity of the fluorescent coating material and secondarily upon the spreading of the electron beam as it penetrates the screen. These problems are easily ignored in TEM today; new coating methods and materials have eliminated the coarse screens of older instruments.

Permanent records, or electron micrographs, are made by exposing photographic film directly to the electron beam, the electrons interacting with the film emulsion in much the same way as light photons in conventional photography. The photographic chamber is located directly beneath the observation screen. (Electron micrography is discussed in Chapter 2.)

VACUUM SYSTEM

The final system of the TEM is the high vacuum system, which is designed to remove all gases from the column. Gas will interact with the electrons and scatter them randomly, giving rise to reduced contrast and/or noise; ionization of gas molecules causes random electrical discharges and fluctuations in the beam. Residual gases will also react with the heated filament, eroding it; and finally gases may contaminate the sample.

The minimum vacuum at which a TEM must be maintained is 10^{-4} Torr because the path length of an electron, or probability of an electron's interacting with a gas molecule, is 2.5 m at that level. A higher vacuum is desirable, and as a rule, the higher the vacuum, the better the conditions for operation. Virtually all modern instruments are designed for operation in the 10^{-6} to 10^{-5} Torr range.

The ideal method of operating a TEM would be to evacuate it, seal it, and switch off the vacuum pumps. Unfortunately, this is impossible; gases are continously entering the system through minute leaks (potentially at every area of the microscope sealed with a gasket) and by the evolution of gas and vapor contaminants from within the column. Sublimation of the specimen during irradiation also causes contamination. The worst offender is photographic film because it accumulates and retains water vapor. For this reason film is usually outgassed at low vacuum prior to loading in the microscope.

The present state of vacuum technology requires that high vacuums be produced in two stages. The stages are arranged in series, each phase being produced by a different type of vacuum pump. The first stage is the production of low vacuum, approximately 10^{-2} Torr, from atmospheric pressure. Another pump takes the vacuum from low to high, or the 10^{-6} to 10^{-4} Torr range. Low vacuum is produced by a mechanical pump, and high vacuum by a diffusion pump.

Rotary Pump

The rotary (synonyms: mechanical, roughing) pump is an oil-immersed, eccentric-vane pump that simply displaces air from one point to another (Figure 1-5). It consists of a cylindrical rotor, bearing a pair of spring-loaded vanes parallel to its axis and mounted inside a cylindrical casing of slightly larger diameter. The rotor is eccentrically mounted so that it touches the casing at only one point giving rise to a line contact. The spring-loaded vanes also bear on the casing in a pair of line contacts. The space between the rotor and casing is thus divided into three compartments, numbered 1, 2, and 3 in Figure 1-5.

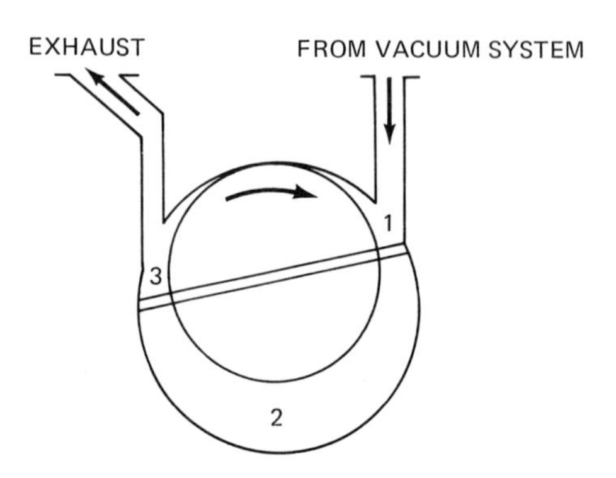

Figure 1-5. Cross section of a rotary pump.

When the rotor revolves, space 1 becomes larger, drawing gas in from the microscope column. Space 2 remains approximately the same size, but 3 becomes smaller. Any gas trapped in space 3 will be forced out the exhaust pipe; as soon as the tip of the vane passes the exhaust, space 3 becomes 1. Thus, there is a continuous cycle of suction, idle, and exhaust.

The highest vacuum that a rotary pump can achieve depends upon the rate at which gas leaks across the vane seals and the rotor-to-casing seal. A rotary pump in good condition will attain 10^{-2} to 10^{-1} Torr; arranged in series, two or more pumps will reach 10^{-4} Torr.

Because of the line contacts, rotary pumps are subject to mechanical wear, which decreases efficiency. Maintenance of the systems is simple, and should be routinely done to ensure optimum performance. Each week check the oil level at the window on the pump, and maintain the recommended level. The oil should be completely removed and replaced annually. Periodically check the fan belt for signs of wear, as abrasion will cause it to break.

Rotary pumps may be located at any distance from the microscope; many are placed in a different room to reduce the annoying noise level and vibrations. The one disadvantage in distant placement is that use of many hose connections may give rise to inefficiency. Consequently, hose connectors should be periodically checked and tightened if necessary.

Diffusion Pump

The high vacuum pump used in series with the rotary is the diffusion pump (Figure 1-6). Containing no mechanical or moving parts, it is something of an engineer's dream. This pump is mounted directly beneath the film chamber, so that it is very efficient (no hoses, etc.). It operates by an ingenious method: Oil or mercury is heated and vaporized at the base of the pump, being forced thereby to move rapidly upward through a tube at the pump's center. As the oil vapor emerges from the top of the tube, the stream is deflected almost 180° by the umbrella. On its downward path the stream traps gas molecules and has sufficient impetus to "throw" gas molecules away from the EM column. This action establishes a pressure

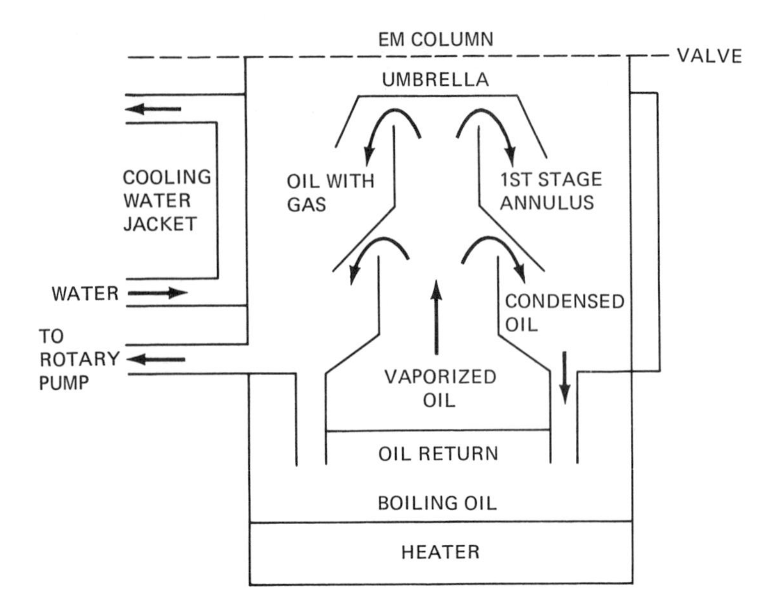

Figure 1-6. A two-stage oil diffusion pump.

difference across the pump umbrella, giving rise to the pumping action. Therefore, gas is free to attain equilibrium by diffusing from the column toward the diffusion pump, but it cannot diffuse back again.

The actual pressure difference depends upon a number of factors, the primary one being the clearance between the umbrella and pump casing. The larger this annular area is made, the more rapidly will the pump reach its ultimate vacuum, and the higher that vacuum will be. However, if the gas pressure in the space below the umbrella rises above a critical value, then the gas that has already been removed will diffuse back into the column across the barrier of moving oil molecules, and the pump will stop operating. Consequently it is necessary to keep the space below the annulus continuously pumped. This is done by having a second-stage umbrella mounted beneath the first. The annular space is smaller so that the concentration of oil molecules is increased, and this stage of the pump can transfer gas from the intermediate pressure region to the space below. This process may be repeated a number of times, the annular space being reduced in area at each step. The limit is reached when the distance between the last umbrella and the casing has reached the reasonable limit of machining tolerance.

When the oil vapor has crossed the annular space of each stage, it strikes the pump casing, which is cooled by water. It then condenses to a liquid and returns by gravity to the boiler, where it is vaporized and recycled. The gas that accumulates below the final stage must be continuously removed by connection to the rotary pump.

In a diffusion pump using mineral oil as the pumping fluid, each stage deals with a pressure difference of about a factor of 10 across it. Four stages will therefore give a total pressure difference of 10^4, or a high vacuum of 10^{-5} Torr if the low vacuum (also called the backing vacuum) pulls 10^{-1} Torr. The highest vacuum is determined by the vapor pressure of the pumping fluid at the temperature of the walls of the condensor. With the oils commonly used, this is in the 10^{-6} Torr range.

A number of precautions must be observed in operating a high vacuum system. First, never expose the hot diffusion pump to the atmosphere. Although most modern microscopes have a fail-safe system to prevent this, neither older instruments nor bell jars do. If this happens, the diffusion pump will be filled with burned oil, more familiarly known as asphalt, and the microscope column or bell jar will be coated with oil. Second, always make sure that the cooling water flow is sufficient; otherwise overheating of the pump will result in these same problems.

In summary, a high vacuum system operates as follows:

1. Low vacuum is achieved from atmosphere with a rotary pump; thus there is a valve between the rotary pump and the TEM.
2. After low vacuum is reached, the warm diffusion pump is opened to the column, and is backed by the rotary pump.
3. To return to atmosphere, the valve is closed between the diffusion pump and the column. Air is admitted while the rotary pump continues to back the diffusion pump.

Vacuum Indicators

Electron microscopes are most commonly equipped with indirect vacuum indicators, which reveal when high vacuum has been reached but do not directly show what that high vacuum is. Although these devices are suitable for routine operation, vacuum leak detection is difficult when they are used.

Low vacuum (1 to 10^{-3} Torr) is commonly measured with a thermocouple gauge. Here, a thermocouple is attached to a heated filament, and a direct relation between temperature and vacuum is established. A pirani gauge will then measure the vacuum from 100 to 10^{-4} Torr; it measures the resistance of a heated filament via a Wheatstone bridge. Ionization gauges measure the range from 10^{-3} to 10^{-10} Torr by ionizing the gas molecules remaining within the system, the number of ions being inversely proportional to the vacuum.

RESOLUTION

The factor that makes electron microscopes powerful tools is their high resolution, resolution being defined as that distance between two adjacent objects at which the objects lose their separate identities. Simple magnification beyond this point is worthless, since one is enlarging an inherently blurred image. Theoretically, the resolving power of the optical microscope is limited by the de Broglie wavelength of light (discussed below), and that of the EM is ultimately controlled by the wavelength of electrons. High resolution requires a small ratio of wavelength to lens diameter; a light microscope uses visible light having a wavelength of several thousand angstroms, whereas an EM's beam is of wavelength less than 1 Å. However, there are four physical factors that severely limit resolution: diffraction, astigmatism, and chromatic and spherical aberrations. For although wavelength ultimately defines resolution, the construction of instruments cannot overcome other optical phenomena.

Diffraction, a phenomenon that determines the maximum resolution obtainable with any type of microscope, is the interference between the component rays of a single broad wave front. Visibly it occurs when light passing near an object is bent into the object where the shadow would normally be. Diffraction has been mathematically proved to be related to wavelength and thus to resolution.

In the nineteenth century, Abbe defined the magnitude of diffraction effects as d, where the term d provides a numerical value for the limit of resolution of any diffraction-free optical system:

$$d = \frac{0.612\lambda}{\eta \sin \alpha}$$

where

d = radius of the first dark ring of an Airy disk measured at the minimum
λ = wavelength of the image-forming radiation
η = index of refraction between the point source and the lens, relative to free space
α = half the angle of the cone of radiation from the specimen plane accepted by the front lens of the objective

Most light microscopists are familiar with the relationship:

$$\eta = \sin \alpha$$

which is usually referred to as the numerical aperture of a lens.

From the Abbe equation it can be derived that to attain maximum resolution with a light microscope, it is necessary to have a minimum d, maximum η, and maximum sin α. Because η cannot be increased beyond approximately 1.5 and α near 90°, the only remaining way to enhance resolution is to decrease λ. As mentioned above, visible light has a wavelength of about 2000 Å, whereas the wavelength of electrons is less than 1 Å.

The French physicist de Broglie advanced the idea that moving particles have wavelike properties. Implicit in this theory is the fact that an electron beam can be used as a type of illumination. de Broglie then concluded that a wavelength can be assigned to moving particles, which may be calculated from the equation:

$$\lambda = \frac{h}{mv}$$

where

λ = wavelength of the particle
h = Planck's constant
= 6.23×10^{-27} erg Sec
m = mass of the particle
v = velocity

When the known values for electrons are substituted, the above expression becomes:

$$\lambda = \frac{12.3}{\sqrt{V}}\ \text{Å}$$

where V = accelerating voltage. Thus the wavelength (and resolution) of a beam of electrons depends upon the potential, V, through which it has been accelerated.

Abbe's equation can then be rewritten, substituting the de Broglie equation for λ:

$$d = \frac{(0.61)(12.3)}{\eta \sin \alpha \sqrt{V}}$$

Because EM angles are always very small, $\sin \alpha \doteq \alpha$, and since both the object and image are in field-free space, the refractive index is $\eta = 1$. The above equation is then rewritten as:

$$d = \frac{7.5}{\alpha \sqrt{V}}\ \text{Å}$$

Therefore, resolution in the EM is ultimately determined by the accelerating voltage and angular aperture of the objective lens. Substituting realistic values in the above equation (voltage = 10^5 V):

$$d = 2.4\ \text{Å}$$

The resolution of the light microscope is:

$$d = \frac{(0.5)(4000)}{(1.5)(\sin 85)}$$

$$= 2000\ \text{Å}$$

This difference in magnitude is responsible for the greater resolving power of the EM.

It must be stressed that the theoretical limitation of resolution as a result of diffraction effects is much better than practical values. Under optimal conditions, typical resolutions are: TEM, 3 Å; STEM (scanning transmission electron microscope), 7 Å; and SEM (scanning electron microscope), 25 Å. Routine operation using a well-prepared sample yields the following resolutions: TEM, 25 Å; STEM, 30 Å; SEM, 60 Å.

Diffraction in the TEM is manifested as a spreading of the beam behind the sample (Figure 1-7). This is not the electron diffraction that is used to identify compounds or elements on the basis of their crystalline nature.

Chromatic aberration, another parameter affecting resolution, arises because the focal length of a lens will vary according to the wavelength of radiation passing through it. In the light microscope it is manifested by light of different colors being focused at different distances from the lens; it has been overcome by using flint glass lenses in conjunction with typical glass lenses.

In the EM, chromatic aberration involves electrons traveling at slightly different velocities (and therefore different wavelengths) due to fluctuations in the accelerating voltage. Consequently, rapidly moving particles pass through a lens's magnetic field quickly without too much influence and are focused farther from the lens than electrons traveling at more moderate speeds (Figure 1-8).

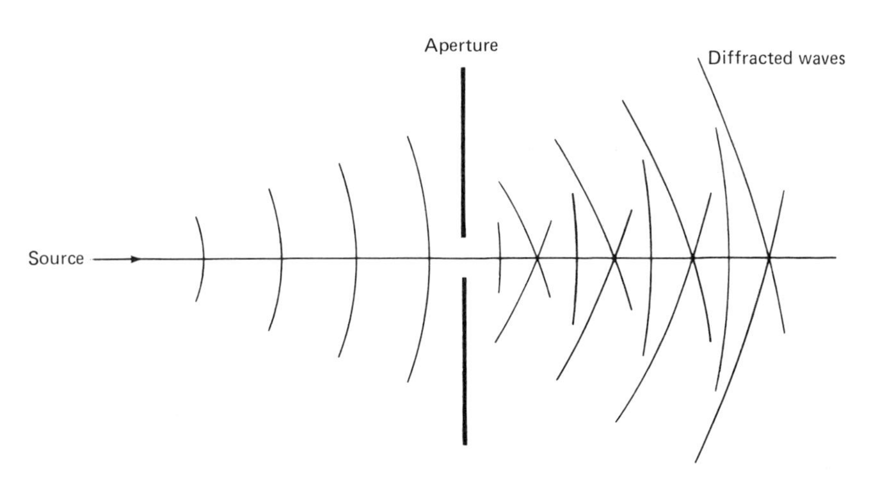

Figure 1-7. Diffraction effects.

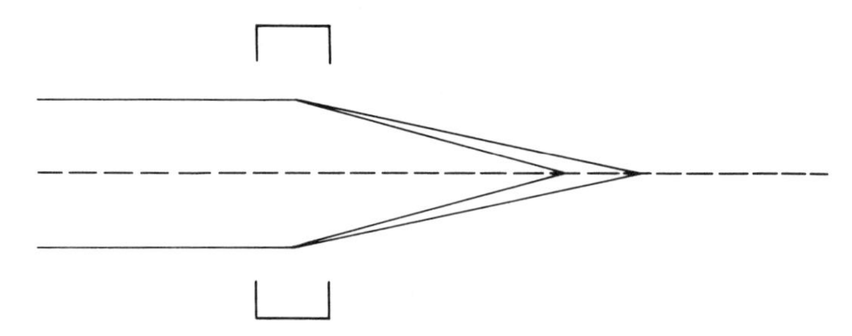

Figure 1-8. Chromatic aberration in electron lenses.

Mathematically, chromatic aberration is related to resolution as follows:

$$d_{cV} = k_c \cdot f \cdot \alpha_o \cdot \frac{\Delta V}{V}$$

$$d_{cI} = 2k_c \cdot \alpha_o \cdot \frac{\Delta I}{I}$$

where

d_{cV} and d_{cI} = separation of two object points which are just resolved, considering voltage and current respectively
k_c = dimensionless constant (0.75–0.1)
f = focal length
α_o = objective aperture angle
V = accelerating voltage
ΔV = maximum departure from V
I = current
ΔI = maximum departure from I

Chromatic aberration is largely eliminated by incorporating a very stable high voltage system in the EM (see below), but it is never completely eliminated because of electron–specimen interaction. In fact, it is largely responsible for the formation of contrast and is thus desirable.

Rewriting the above equation will demonstrate the maximum fluctuation tolerable in a high voltage (HV) system:

$$\frac{\Delta V}{V} = \frac{d_{cV}}{k_c \cdot f \cdot \alpha}$$

If we substitute realistic values:

d_{cV} = 10 Å; the maximum resolution resulting from diffraction and spherical aberration
$k_c = 0.75$
$\alpha = 4.5 \times 10^{-3}$ radians (optimum aperture angle)
$f = 3$ mm

then:

$$\frac{\Delta V}{V} = \frac{10 \text{ Å}}{(0.75)(4.5 \times 10^{-3} \text{ radians})(3 \text{ mm})}$$

$$= 1.0 \times 10^{-4} \text{ of the voltage}$$

One may also calculate the variation in lens current as related to chromatic aberration:

$$d_{cI} = 2k_c \cdot f \cdot \alpha \frac{\Delta I}{I}$$

$$\frac{\Delta I}{I} = 5 \times 10^{-5} \text{ of the current}$$

The third optical parameter influencing resolution is spherical aberration, which depends upon the aperture angle of the incident radiation. The most serious geometric aberration encountered in electron optics, it occurs when electrons leaving the object and passing close to the center of the objective lens are focused in one plane, while electrons passing through the outer edge of the magnetic field are focused in another plane. Electrons passing through the periphery have a shorter focal length because the magnetic field is always much stronger at the perimeter. (See Figure 1-9).

The only way to reduce spherical aberration is to introduce an aperture just beneath the objective lens. It allows only those elec-

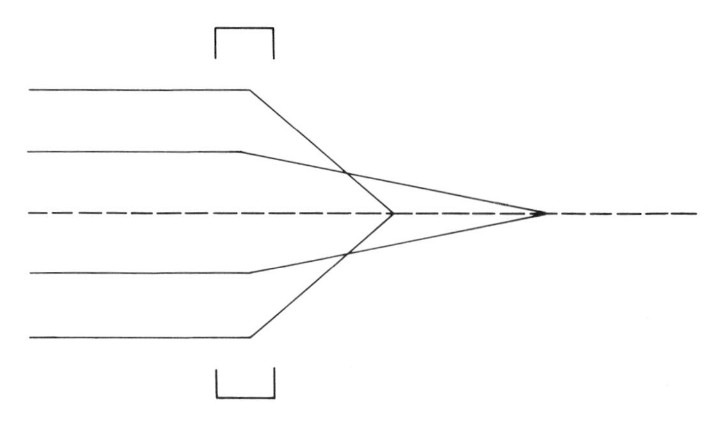

Figure 1-9. Spherical aberration in electron lenses.

trons passing through the optical axis to form the final image; peripheral electrons are stopped by the aperture. Unfortunately, this aperture severely reduces resolution because it limits the numerical aperture of the lens. That is:

$$d_s = k_s \cdot f \cdot \alpha_o{}^3$$

where

d_s = separation of two object points which are just resolved
k_s = dimensionless proportionality constant
f = focal length
α_o = objective aperture angle

The final optical limitation influencing resolution is astigmatism, which develops when a lens focuses more strongly along one axis than along another owing to minute flaws or inhomogeneities within the coilings (Figure 1-10). As a result, the image can never be clearly focused. It is corrected by superimposing on the objective lens magnetic field another field having a deliberate asymmetric distribution and variable magnitude; it is then positioned so as to oppose and cancel the existing lens's asymmetry. The correction is made by a stigmator, which functions as an additional weak cylindrical lens of strength just sufficient to correct the cylindrical component of the objective.

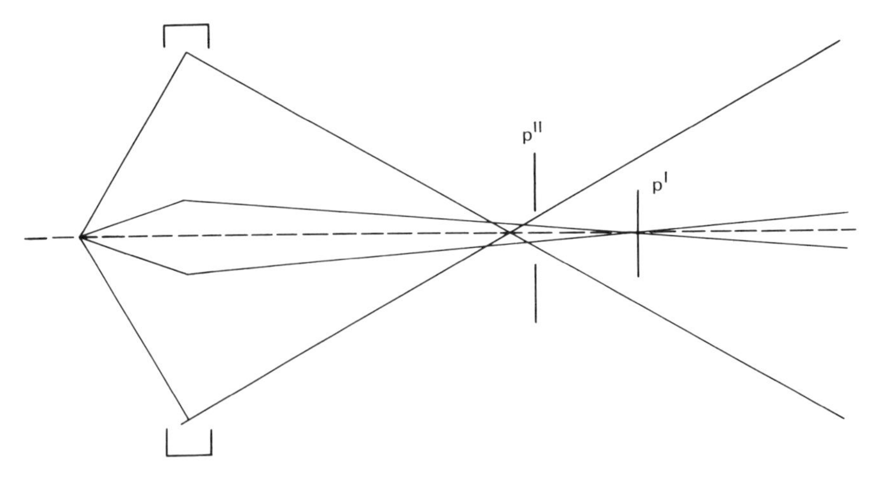

Figure 1-10. Astigmatism.

Astigmatism in the TEM is manifested by a change in shape, for example, of a round object into an oval, as focus is varied. Unless it is extremely pronounced, it becomes apparent in the TEM at magnification greater than 10,000X. Therefore, it should be corrected at high magnification by "focusing" with the stigmator until the phenomenon is eliminated. An image is not stigmatic when a shape remains the same as the focus is varied. That is, a circle symmetrically expands out of focus and back into focus; it does not go from a circle into an oval as focus is altered. All samples must be checked for astigmatism.

In summary, diffraction and spherical aberration effects are beyond operator control, whereas chromatic aberration and astigmatism are influenced by proper operation.

THE TRANSMISSION ELECTRON MICROSCOPE: INSTRUMENTATION BIBLIOGRAPHY

Agar, A. W. (1967) The operation of the electron microscope. In: *Techniques for Electron Microscopy*, 2nd ed., p. 1. Kay, D. H. (ed.). Blackwell Scientific Pub., Oxford.

——, R. H. Alderson, and D. Chescoe. (1974) Principles and practice of electron microscope operation. In: *Practical Methods in Electron Microscopy*, vol. 2 Glauert, A. M. (ed.). American Elsevier Pub. Co., New York.

Cosslett, V. E. (1946) *Introduction to Electron Optics.* Clarendon Press, Oxford.
—— (1969) Energy loss and chromatic aberration in electron microscopy. *Z. Angew. Phys.* 27:138.
—— (1971) Electron microscopy. In: *Physical Techniques in Biological Research*, 2nd ed., 1A:71. Oster, G. (ed.). Academic Press, New York.
Cox, R. W., and R. W. Horne. (1968) Accurate calibration of the magnification of the A. E. I. E. M. 6B/2 electron microscope using catalase crystals. *Proc. 4th Eur. Reg. Conf. EM (Rome)* 1:579.
Crick, R. A., and D. L. Misell. (1971) A theoretical consideration of some defects in electron optical images. A formulation of the problem for the incoherent case. *J. Appl. Phys.* 4:1.
Cundy, S. L., A. Howie, and U. Valdre. (1969) Preservation of electron microscope image contrast after inelastic scattering. *Phil. Mag.* 20:147.
Grivet, P. (1972) *Electron Optics*, 2nd ed. Pergamon Press, New York.
Guthrie, A. (1963) *Vacuum Technology.* John Wiley and Sons, New York.
Hall, P. M. (1968) Vacuum technology. In: *Thin Film Technology*, p. 19. Berry, R. W., P. M. Hall, and M. T. Harris (eds.). Van Nostrand Reinhold, New York.
Heidenreich, R. D., W. M. Hess, and L. I. Ban. (1968) A test object and criteria for high resolution microscopy. *J. Appl. Crystallog.* 1:1.
Lipson, S. G., and H. Lipson. (1969) *Optical Physics.* Cambridge Univ. Press, Cambridge.
Longhurst, R. S. (1967) *Geometrical and Physical Optics.* John Wiley and Sons, New York.
Parsons, D. F. (1970) Problems in high resolution electron microscopy of biological materials in their natural state. In: *Some Biological Techniques in Electron Microscopy*, p. 1. Parsons, D. F. (ed.). Academic Press, New York.
——, and P. A. Redhead. (1962) Ultrahigh vacuum. *Sci. Am.* 216:1.
Wischnitzer, S. (1970) *Introduction to Electron Microscopy*, 2nd ed. Pergamon Press, New York.
——, (1973) The electron microscope. In: *Principles and Techniques of Electron Microscopy* 3:3. Hayat, M. A. (ed.). Van Nostrand Reinhold, New York.
Wrigley, N. G. (1968) The lattice spacing of crystalline catalase as an internal standard of length in electron microscopy. *J. Ultrastr. Res.* 24:454.

SELECTED GENERAL ELECTRON MICROSCOPY BIBLIOGRAPHY

Andrews, K. W., D. J. Dyson, and S. R. Keown. (1967) *Interpretation of Electron Diffraction Patterns.* Hilger and Watts, Ltd., London.
Barer, R., and V. E. Cosslett. *Advanced Optical and Electron Microscopy.* Academic Press, New York. Multivolume series.
Beeston, B. E. P., R. W. Horne, and R. Markham. (1973) Electron diffraction and optical diffraction techniques. In: *Practical Methods in Electron Microscopy*, vol. 1, part 2. Glauert, A. M. (ed.). American Elsevier Publishing Co., New York.

Bevelander, G. (1970) *Essentials of Histology.* C. V. Mosby Co., St. Louis.

Birks, L. S. (1972) *Electron Probe Microanalysis.* Wiley-Interscience, New York.

Dawes, C. J. (1971) *Biological Techniques in Electron Microscopy.* Barnes and Noble, Inc., New York.

DeHoff, R. T., and F. N. Rhines. (1968) *Quantitative Microscopy.* McGraw Hill Book Co., New York.

Fawcett, D. W. (1966) *The Cell: Its Organelles and Inclusions.* W. B. Saunders, Philadelphia.

Frey-Wyssling, A., and K. Mühlethaler. (1965) *Ultrastructural Plant Cytology.* American Elsevier Publishing Co., New York.

Glauert, A. M. (ed.). (1974–76) *Practical Methods in Electron Microscopy.* American Elsevier Publishing Co., New York.

Goodhew, P. R. (1974) Specimen preparation in materials science. In: *Practical Methods in Electron Microscopy,* vol. 1, part 1. Glauert, A. M. (ed.). American Elsevier Publishing Co., New York.

Grimstone, A. V. (1968) *The Electron Microscope in Biology.* Arnold, London.

Harris, R. J. C. (ed.). *The Interpretation of Ultrastructure.* Academic Press, New York. Multivolume series.

Hayat, M. A. (1972) *Basic Electron Microscopy Techniques.* Van Nostrand Reinhold Co., New York.

—— (ed.). *Principles and Techniques of Electron Microscopy: Biological Applications.* Van Nostrand Reinhold Co., New York. Vol. 1: 1970. Vol. 2: 1972. Vol. 3: 1973. Vol. 4: 1974. Vol. 5: 1975. Vol. 6: 1976. Vol. 7: 1977.

—— (ed.). *Electron Microscopy of Enzymes: Principles and Methods.* Van Nostrand Reinhold Co., New York. Vol. 1: 1973. Vol. 2: 1974. Vol. 3: 1974. Vol. 4: 1975.

Juniper, B. E., G. C. Cos, A. J. Gilchrist, and P. R. Williams. (1970) *Techniques for Plant Electron Microscopy.* Blackwell Scientific Pub., Oxford.

Kay, D. (ed.). (1967) *Techniques for Electron Microscopy,* 2nd ed. Blackwell Scientific Pub., Oxford.

Koehler, J. K. (1972) *Electron Microscope Techniques in Biology.* Springer-Verlag, Berlin.

—— (ed.). (1973) *Biological Electron Microscopy.* Springer-Verlag, New York.

Magnan, C. (ed.). (1961) *Traité de Microscopie Electronique,* 2 vol. Herrman, Paris.

McGee-Russel, S. M., and K. F. A. Ross (eds.). (1968) *The Interpretation of Cell Structure.* Edward Arnold, Ltd., London.

McKinley, T. D., K. E. J. Heinrich, and D. B. Wittry (eds.). (1966) *The Electron Microprobe.* John Wiley and Sons, New York.

Meek, G. A. (1970) *Practical Electron Microscopy for Biologists.* Wiley-Interscience, New York.

Mercer, E. H., and M. S. C. Birbeck. (1966) *Electron Microscopy: A Handbook for Biologists,* 2nd ed. Blackwell Scientific Pub., Oxford.

Oster, G. (ed.). (1971) *Physical Techniques in Biological Research,* 2nd ed. Academic Press, New York.

Parsons, D. F. (ed.). (1970) *Some Biological Techniques in Electron Microscopy.* Academic Press, New York.
Rogers, A. W. (1967) *Techniques of Autoradiography.* American Elsevier Publishing Co., New York.
Roth, L. J., and W. E. Stumpf (eds.). (1969) *Autoradiography of Diffusible Substances*, Academic Press, New York.
Seigel, B. (ed.). (1964) *Modern Developments in Electron Microscopy.* Academic Press, New York.
Sjostrand, F. S. (1967) *Electron Microscopy of Cells and Tissues. Instrumentation and Techniques.* Academic Press, New York.
Thomas, G. (ed.). (1972) *Electron Microscopy and Structures of Materials.* Univ. of California Press, Berkeley.
—— and M. J. Goringe. (1980) *Transmission Electron Microscopy of Materials.* John Wiley and Sons, New York.
Wischnitzer, S. (1970) *Introduction to Electron Microscopy*, 2nd ed. Pergamon Press, Elsmford, N.Y.

2. Electron Micrography

Permanent records of electron images are made by directly exposing film to the electron image. The film will exactly reproduce the information contained in the electron beam, but in negative contrast, resulting in an electron micrograph. Conventional photographic processing by wet chemical means, enlarging, and printing the negative produces a positive print. The major differences between electron micrographs and light photographs are that roughly one incident electron is the equivalent of 10 to 100 light photons, and electron micrographs are processed in a manner that will enhance contrast. After removal of the electron-sensitive film from the microscope, the processing is identical to that of light-sensitive film.

Various authors have discussed the theoretical aspects of electron interactions with film emulsions (Valentine, 1966; Hamilton and Marchant, 1967; Burge and Garrard, 1968; Burge et al., 1968; Iwanaga et al., 1968; Farnell and Flint, 1973, 1975). The following discussion deals with practical aspects of electron micrography, and the reader should consult the above references for theoretical data (which are not introduced here because dependable and readily available films for recording electron micrographs are marketed).

One area in electron micrography that is still being developed is that of recording the image outside of the vacuum chamber (Guetter and Menzel, 1978). Recall that the film is held in a chamber directly beneath the observation screen; unless it is prepumped in a vacuum chamber to remove water vapor and other volatile materials, the film will outgas in the TEM, which leads to contamination of the microscope column and requires long evacuation schedules that essentially are down-time for the instrument. Kumpf (1980) introduced a fiber optics system that records electron micrographs outside of the microscope proper. This system and others are being developed to avoid the problems of vacuum-held film and make electron micrography more efficient.

IMAGE FORMATION

The film used for recording images is basically an emulsion supported on a polyester sheet (~0.18 mm thick) or glass plate. The emulsion (12–25 μm thick) consists of small crystals of silver halide (0.5–10 μm in diameter), also referred to as grains, homogeneously dispersed and held in place by gelatin. The glass plate or polyester sheet simply serves as a support for the emulsion, and these materials also do not distort during vacuum exposure (i.e., they do not curl or shrink). Prior to the introduction of polyester substrates, glass was conventionally used because other materials were unstable. Although glass is indeed more stable than polyester (which may undergo slight dimensional changes), it is unwieldy, and breakage will destroy the micrograph. As a result, polyester-supported emulsions are routinely used except when ultrahigh resolution and/or image processing is necessary; the last two situations are encountered only in isolated research experiments.

When an electron enters an emulsion, it is probable that it will strike and impart some or all of its energy to a silver halide crystal. These excited electrons are trapped, and in the presence of silver ions (which naturally exist in emulsions) give rise to aggregates of silver atoms, referred to as latent image specks, which will form the negative image when chemically processed (developed to metallic silver) (Hamilton and Urbach, 1966). The films used in electron microscopy are fine-grain, meaning that the probability of an image-forming electron's being recorded is high.

Electrons that interact with the emulsion travel at different speeds because image formation is by interaction of the primary beam with the specimen (elastic and inelastic scatterings). Some electrons will travel at a slower rate than others because they have been scattered by electron-dense sites within the specimen; those electrons not scattered by the specimen (an electron-transparent region) travel at

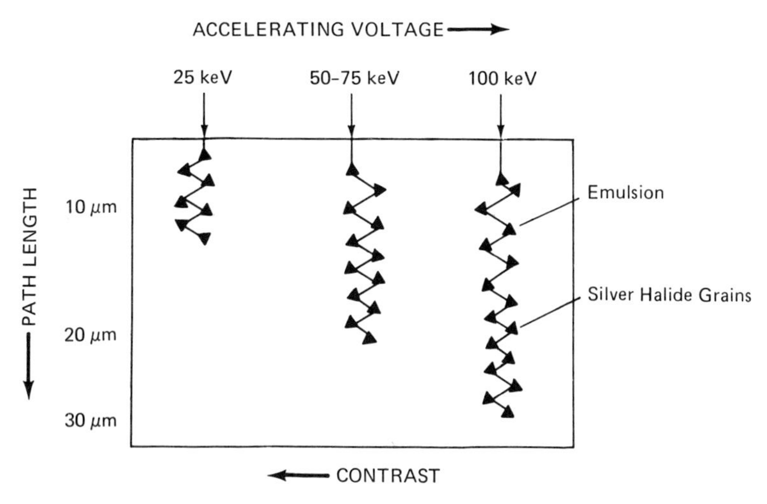

Figure 2-1. The influence of accelerating voltage on path length in an emulsion.

a speed near that of the accelerating voltage. As a result, some electrons carry enough energy to have multiple interactions (Frieser and Klein, 1958) with the silver halide crystals and penetrate deep within the emulsion. Commonly referred to as path length, this event is more exaggerated when different accelerating voltages are compared (Figure 2-1). The path length proportionately increases with accelerating voltage, or in other terms, the stopping power of the emulsion is inversely proportional to higher voltages (Burge and Garrard, 1968; Burge et al., 1968; Herz, 1969). This is reflected in the varying degrees of contrast observed at different accelerating voltages: At 25 keV (Figure 2-2) the image is confined to the uppermost layer of the emulsion and contrast is high; moderate voltages (50–75 keV) produce moderate contrast (Figures 2-3 and 2-4); and higher voltage (100 keV; Figure 2-5) results in low contrast (Burge et al., 1968). Contrast is the ratio of minimal to maximal optical density in an emulsion, and a compromise between degree of contrast and desired resolution must be made. Contrast may be enhanced by using a small objective aperture when working at higher voltages, or by exposing the film to the beam for an optimal duration (i.e., contrast is directly proportional to exposure time).

Also note in Figure 2-1 that as voltage increases, an electron will penetrate deeper before interacting with a grain. The spread function

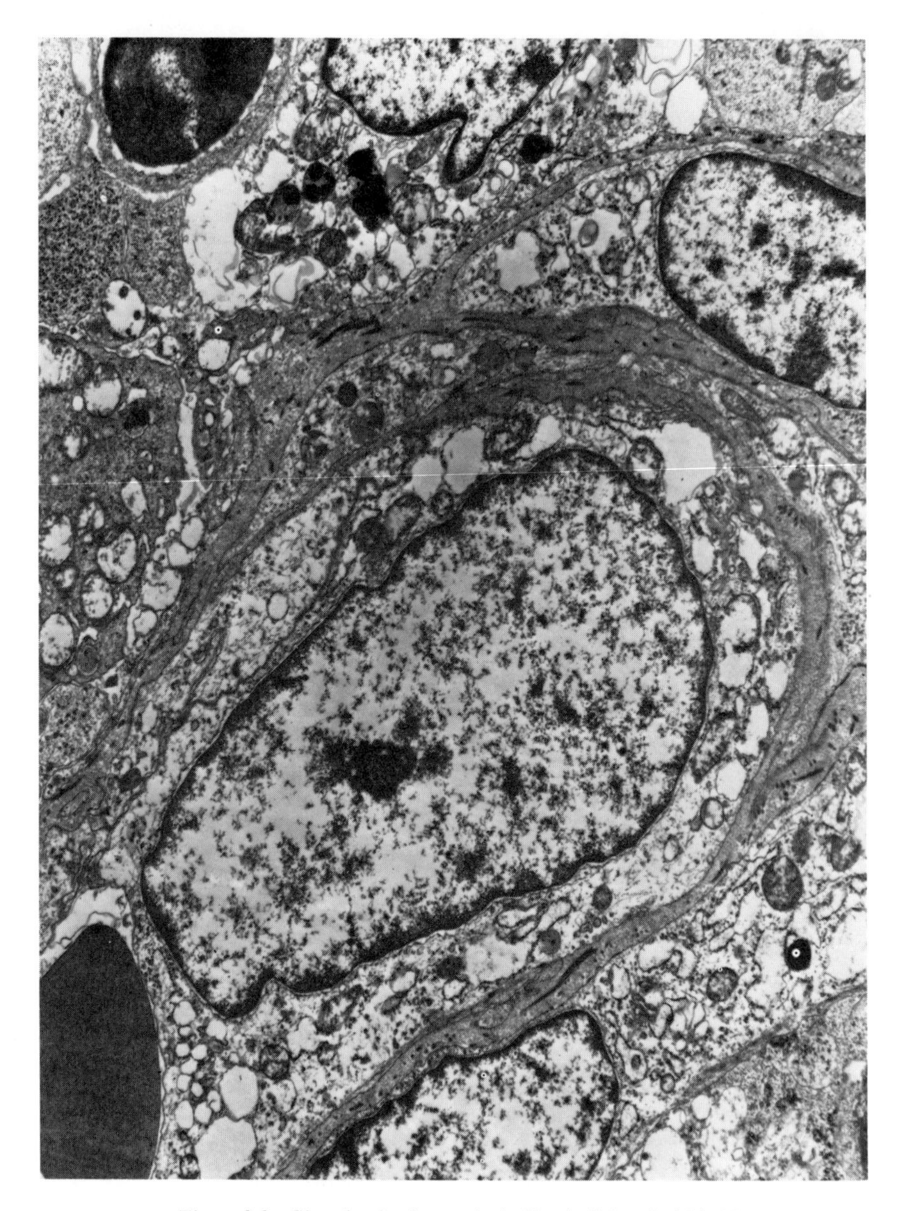

Figure 2-2. Giant benign hyperplasia 25 μA, C-2 at 2, 25 keV.

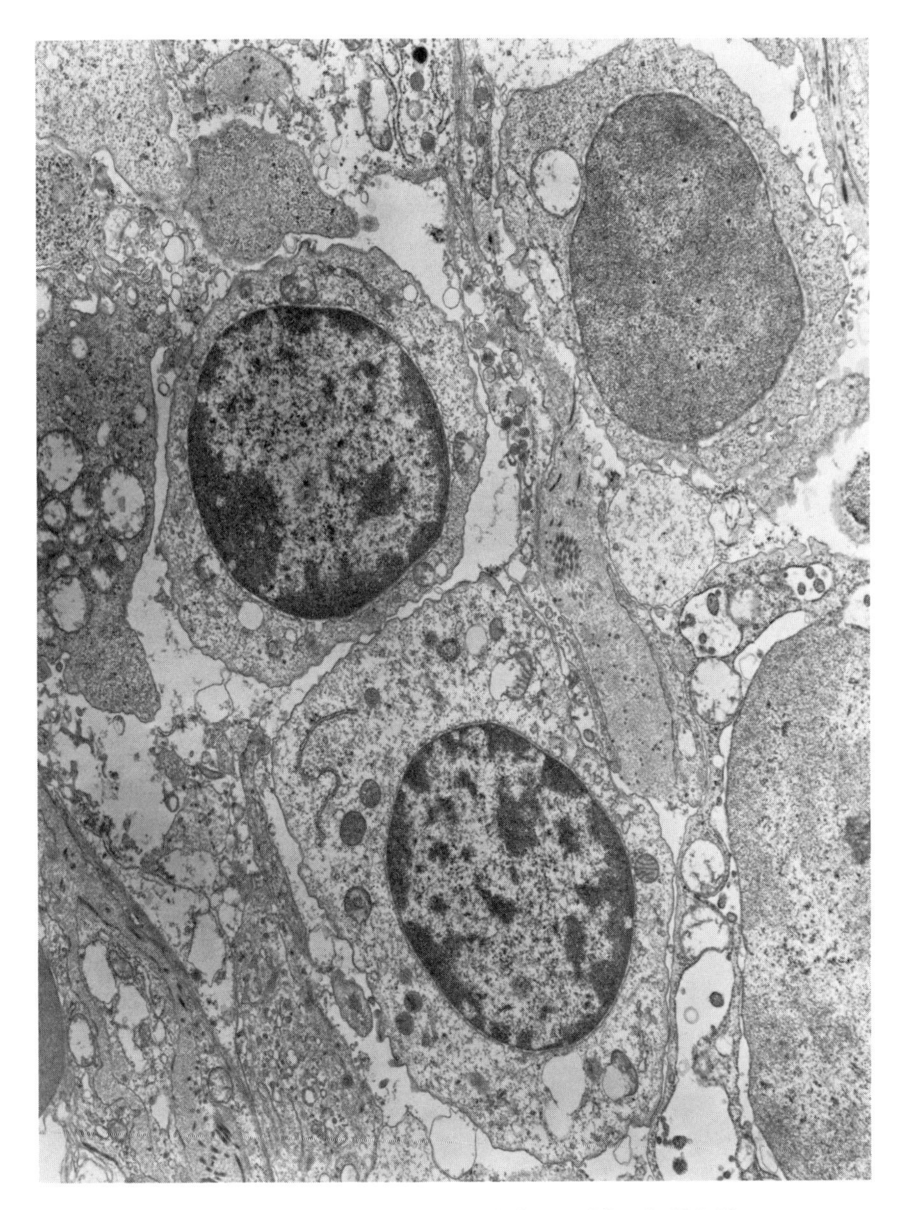

Figure 2-3. Giant benign hyperplasia 25 μA, C-2 at 2, 50 keV.

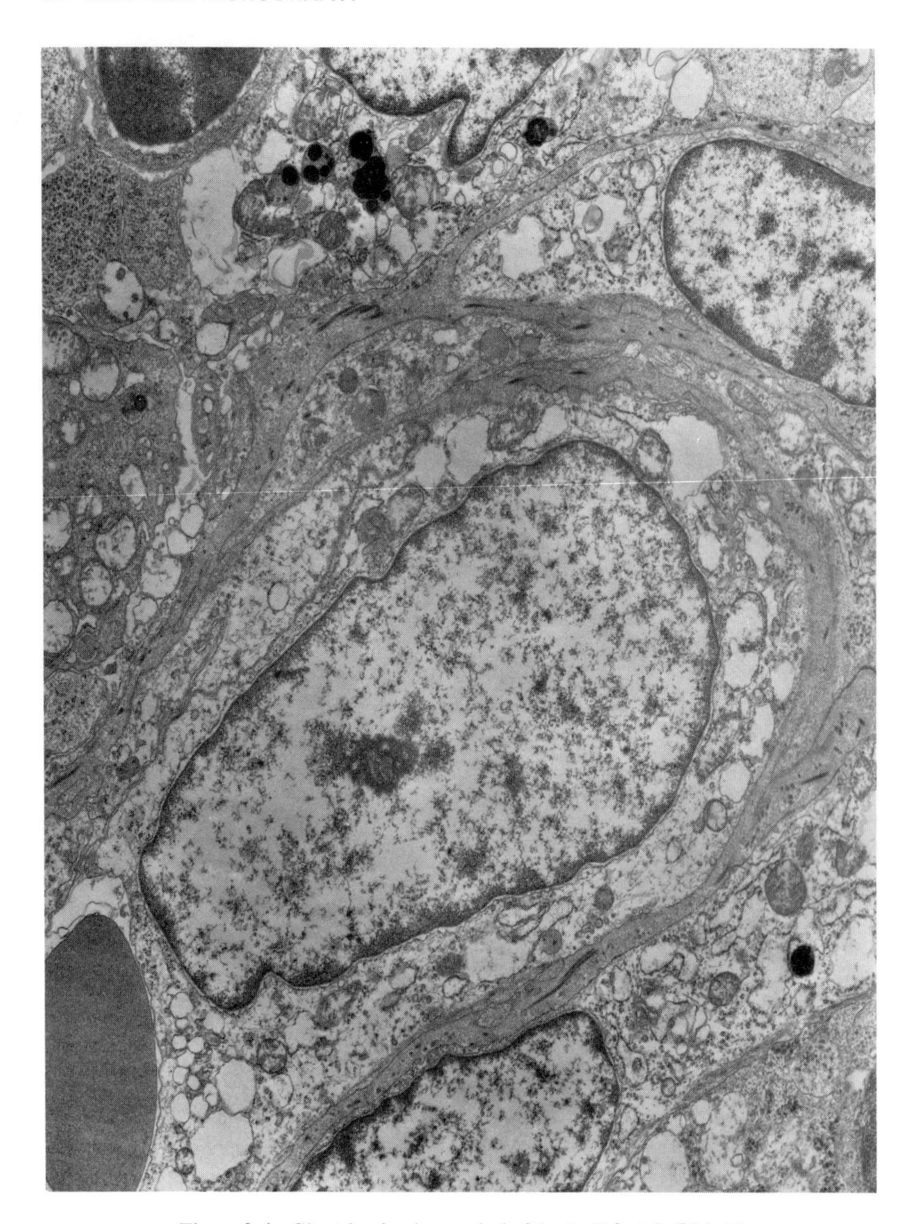

Figure 2-4. Giant benign hyperplasia 25 μA, C-2 at 2, 75 keV.

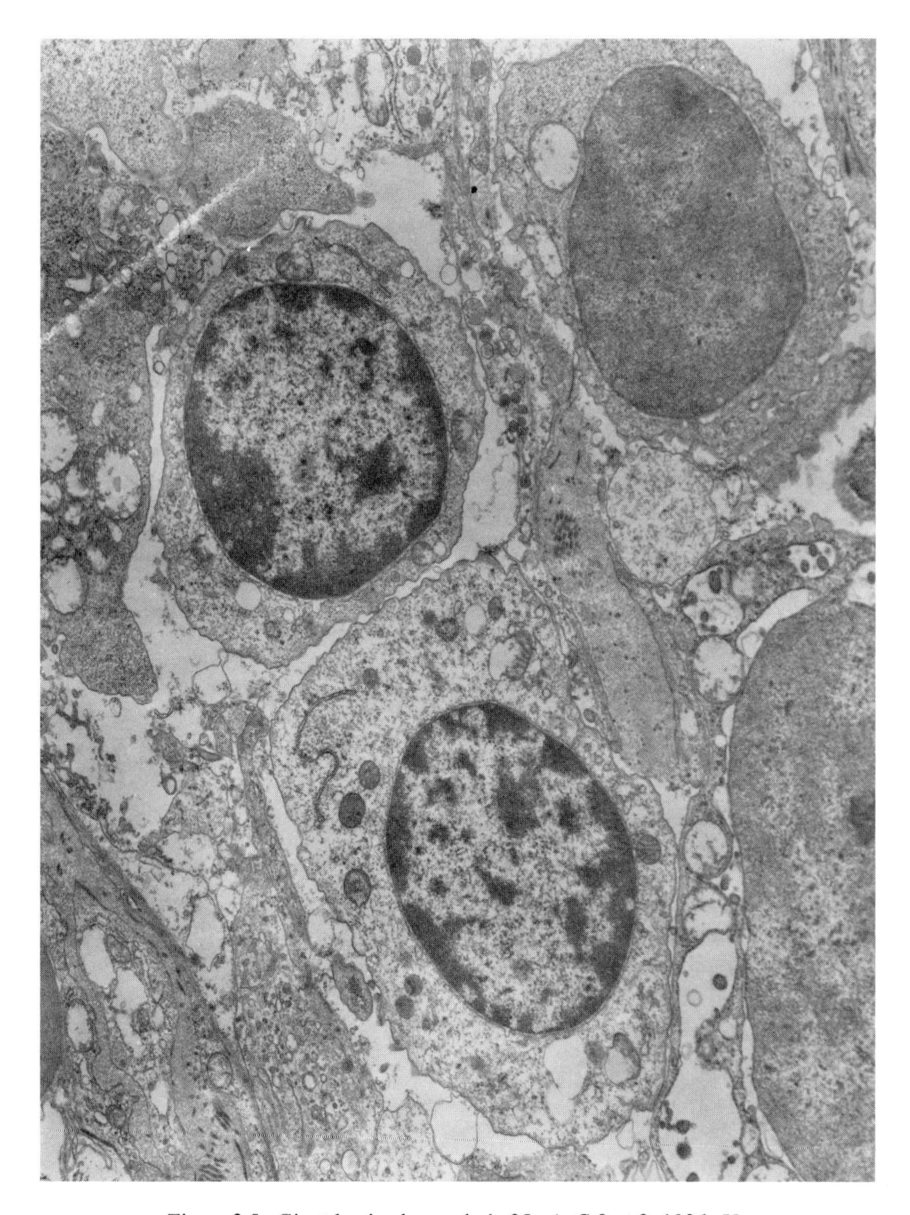

Figure 2-5. Giant benign hyperplasia 25 μA, C-2 at 2, 100 keV.

of an emulsion is the volume of gelatin traversed by the electron, and it increases with accelerating voltage. Eastman Kodak (1973a) estimated the spread function of EM film as 5–10 μm; the practical effect of this is that it limits subsequent enlargement to about 20 diameters.

The signal-to-noise ratio (SNR) refers to the quality of the information being recorded. A high SNR means that more information is being collected, usually by increasing the duration or intensity of exposure. Unfortunately, most biological specimens are unstable if very long exposures are used; typically, short-duration (2–6 sec) exposures are necessary for a good SNR. Under other conditions (e.g., for recording electron diffraction patterns), if radiation damage mechanisms such as heat damage are eliminated, lengthy (60 sec or longer) exposures are possible. Another method for increasing the SNR is to work with small-diameter objective apertures, which will reduce the amount of noise in the beam. A completely different option is to change the development rate or concentration (Farnell and Flint, 1969, 1970). Under routine working conditions, however, the microscopist should use development as a control, and vary contrast effects with the microscope (e.g., use different accelerating voltages or apertures). Contrast may be enhanced during positive printing with high numerical grade papers.

PROCESSING

The wet processing of negatives and positive printing paper follows this sequence: development, stop bath, fixation, washing, and drying. This progression is identical to that of conventional light photography; thus almost any literature dealing with film processing may be consulted (e.g., Adams, 1968; Jacobson and Jacobson, 1972; Kirkpatrick, 1975; Eastman Kodak, 1976a,b; Neblette, 1976). Eastman Kodak (1973a) offers a convenient summary on processing of electron micrographs, and Farnell and Flint (1975) thoroughly discuss both theoretical and practical considerations of electron recordings.

Film must always be carefully handled, as scratches, fingerprints, and dust will severely detract from an image. Damage may occur prior to exposure as well as during chemical processing. Handle the film only by its edges, and wear lint-free gloves, if desired. In addition, know the safelight conditions for the film (i.e., whether amber or red illumination is safe for the specific film employed). This information is supplied by the manufacturer, as are the ambient processing conditions (temperature, time, concentration). The types of film used for recording are, for example, Ilford Electron Microscope Film and Kodak Electron Image Film. The manufacturers also supply developers and fixers suitable for their film.

Development is chemical reduction of activated silver halide grains to black metallic silver under alkaline conditions. Only those grains formed by interaction with the electron beam are reduced under optimal conditions–proper temperature, duration, and agitation during development; an increase of any of these variables may result in fogging, due to attack of nonactivated grains by developer, whereas a decrease results in incomplete reduction.

A dilute solution of a benzene derivative (e.g., methyl-*p*-aminophenol sulfate; Eastman Kodak, 1973b, 1976a) in water constitutes a developer. The water acts both as a solvent and to soften the gelatin emulsion, thus permitting the actual reducing agent to react with the grains. Alkaline conditions are created by activators, which catalyze reduction. To avoid rapid oxidation of the solutions by interaction with air, a low concentration of sodium sulfite functions as a preservative. Fresh developing solutions are pale yellow, but gold when oxidized; discard exhausted solutions. Stock solutions are kept tightly covered in dark bottles; when the working solution is prepared, the stock may be kept reactive by displacing the air with a nonreactive material (e.g., make up the volume by placing glass marbles in the stock bottle). Because working solutions are exposed to the atmosphere and will become exhausted with use, they show oxidative effects more rapidly than stock solutions.

In reality, commercially available developing reagents contain all of the above chemicals, and simple dissolution in a given volume of water and at specified temperature is required for preparation. All processing reagents are prepared in a room separate from the one where film is handled; neither the darkroom nor the microscope lab is appropriate. The powdered chemicals may become suspended in air, and if they settle on film, will begin to react. Consequently, prepare the reagents in a distant lab. Practical examples of commercially available developers are Kodak D-19, a high-contrast negative developer, and Kodak Dektol, for printing papers.

The duration, temperature, and degree of agitation control the development activity, uniform results requiring that these factors be held constant. Each manufacturer specifies the ambient conditions. The duration and temperature are usually 4 min. at 20°C. Agitation accelerates contact between the developer and emulsion, and helps to displace exhausted developer within the emulsion. Agitation is by hand or, if available, with a nitrogen-burst agitation system (Eastman Kodak, 1973b); it is most critical during the first minute or so of development, but negatives or papers should be periodically agitated throughout processing. When the film is initially submerged in the developer (a tank for negatives and a tray for printing paper), agitate it vigorously to dislodge air bubbles at the film surface; trapped bubbles prevent contact of the underlying region by the developer.

Underdeveloped negatives or prints exhibit streaks, low contrast, or uneven density from area to area. These problems result when exhausted or incompletely dissolved developer is used, there is insufficient agitation during processing, or the development time is too short. Overdevelopment increases the general density of the negative, a result of warm or prolonged development. If there has been proper development but the negatives or prints still show poor contrast and the edges of the film are dark, a light leak (e.g., from poor safelight conditions) may have damaged the unexposed film. Another precaution when developing papers involves observation of the image as it emerges. Many novices evaluate the degree of contrast and promptly remove the print assuming that it will overdevelop. What these individuals are forgetting is that this observation is under amber illumination and therefore appears darker than under conventional white lighting. Also, the very rapid appearance of dark areas of the image correspond to those areas containing the highest proportion of latent image specks; the emergence of the fine-detail, light-gray regions (lower proportion of latent image specks) is proportionately slower.

The second step in processing film is treatment with a stop bath, which removes excess, unreacted developer from the film (it stops the action of the developer), and prevents rapid exhaustion of the final fixing solution. Negatives are simply rinsed in cool water for 1–1.5 min., whereas papers are submerged in a weak acetic acid stop bath for a recommended time. Water stop baths must be cool because the emulsion gelation is softened by water. Again agitate to thoroughly replace the developer with stop bath. Acid stop baths may contain an indicator dye to reveal solution exhaustion. Insufficient action of the stop bath induces a mottled appearance.

The unactivated silver halide grains (i.e., those that did not interact with the electron beam) are removed during fixation; these areas correspond to transparent regions on negatives. A fixer is an aqueous solution of sodium thiosulfate ("hypo") and functions by converting silver halide into silver thiocyanate, which is water-soluble. Fixation is under acidic conditions, which will neutralize excess, alkaline developer, thus preventing image fogging. Again, agitation is necessary for thorough infiltration of the emulsion by the fixer; the fixation time and temperature being ~10 min. at 20°C. With fixer use the concentration of silver salts will increase, with formation of an insoluble precipitate. Various "hypo-checks" are available, and the solutions should be checked before use. An exhausted fixer causes yellow staining and excessively soft gelatin.

The film is then rinsed in running water to clear the emulsion and prevent fading over time. Washing must be thorough to prevent crystallization of the fixer both in and at the surface of the emulsion. At this stage room lights may be turned on. The water must be cool (≤20°C) to avoid excessive softening of the emulsion. Negatives are rinsed for ~20 min., whereas the duration of print washing is largely a function of the emulsion support; paper-based prints require significantly longer washing (30–60 min.) than do water-resistant resin-coated bases (4 min.). For this reason resin-coated papers are preferred in routine laboratory situations, but are not suitable for publication purposes because the surface is easily scratched.

After washing the negative is dipped in a dilute wetting agent to prevent waterspots, and finally air-dried under cover. Paper-based prints are dried in a drying drum, while resin-coated papers are air-dried. Negatives must be carefully handled to avoid scratches and dust contamination; store them individually in envelopes.

PRINTING

The final stage in producing an electron micrograph involves positive printing of the negative image. Conventional black-and-white photographic methods are employed (see, for example, Hillson, 1969; Carroll, 1974; Wall and Jordan, 1974; Kirkpatrick, 1975; Eastman Kodak, 1976b). All equipment (e.g., enlarger, timers, processing trays, etc.) and expendable materials (chemicals and papers) may be purchased from local photographic suppliers. For this reason, only general information is presented here.

Numerous types of glossy printing paper are commercially available, the two general classifications being papers that are paper-based and those that are resin-coated. Examples of the former are Agfa Gaevaert, Ilford, and Kodabromide papers, while Kodak Polycontrast Resin Coated Paper is an example of the latter. Paper-based papers will become saturated with the processing solutions and consequently require significantly longer durations of each stage, especially washing. In comparison, only the emulsion of resin-coated papers takes up each solution, and processing is very rapid. Paper-based print paper is available in different contrast grades of 0 to 5, with 5, or high-contrast paper, being most commonly used in EM. In comparison, the contrast of polycontrast papers is controlled by altering the color of the exposing light with graded filters; seven gelatin filters, ranging in color from pale amber (#1 filter, low contrast) to magenta (#4 filter, high contrast), make it possible to effectively change the paper grade without changing paper types (Eastman Kodak, 1973d).

Resin-coated papers are most amenable to routine laboratory situations because a minimal processing time is necessary; on the other hand, they are comparatively more expensive. Paper-based printing paper is preferred, for example, for publication; resins will scratch, and if prints are to be evaluated by several reviewers, the print may be damaged. A grade 4 or 5 glossy print should be used. Information on specific papers may be obtained from photographic suppliers or directly from the manufacturer.

The basic procedure for enlarging a negative and tray-processing a print is as follows:

1. Mount the negative, emulsion side down, in the enlarger negative carrier.
   a. The focal length of the enlarger lens should equal or exceed the diagonal of the film format employed: for example, plates $3\frac{1}{4}'' \times 4''$ require a 135-mm enlarging lens.
   b. Moderate f-stops enhance contrast.

2. Focus the image by adjusting the enlarger bellows. A useful tool is an image (grain) focuser, which will magnify the image ten diameters and thus facilitate good focusing.
3. Adjust the printing easel so that the negative image fills the area where paper will be placed.
4. Determine the proper exposure time by exposing and processing a test strip. This is not necessary for each and every negative when proper exposure conditions (brightness and time) are used in the microscope; usually, only fine adjustments are then required in printing.
   a. A test exposure is made by placing an unexposed strip of paper in the easel (enlarger off), and on top of this the Kodak Projector Print Scale (a film divided into wedges having corresponding exposure time and varying in tone from transparent to opaque). Expose the paper for 1 min. and develop it by standard methods (see below).
   b. The wedge showing highest contrast and still preserving details indicates the proper exposure time.
   c. Do not alter enlarger conditions; if higher contrast is desired, consult data sheets for changing conditions (e.g., *Kodak Data Guide*).
5. Place a new sheet of paper in the easel, and expose it for the desired length of time.
6. Develop the paper with agitation according to manufacturer specifications.
7. Transfer it to stop bath, and agitate the solution.
8. Carry it into the fixer and process it according to manufacturer specifications.
9. Wash in cool running water, 1 hr for paper base or 5 min. for resin-coated stocks.
10. Resin-coated papers are air-dried, while paper-based prints are usually dried on a drying drum.

Aesthetically pleasing micrographs are characterized by equal margins, even exposure across the print (i.e., no flare spots resulting from misalignment of the enlarger's condenser lenses), and absence of fingerprints or scratches. Any flaw seriously degrades the appeal of an otherwise good micrograph.

Other problems are traceable to operating conditions in the microscope. A properly aligned column corrected for astigmatism is essential for good micrography. Any aberration present in the microscope will be reproduced on the micrograph. Misalignment results in uneven illumination and low contrast; an unfocused image may be caused by misalignment or astigmatism. To avoid the frustration of discovering these problems after printing, a microscope should always be maintained in its optimal operating state.

REFERENCES

Adams, A. (1968) *The Negative.* Morgan Press, Hastings-on-Hudson, N.Y.

Burge, R. E., and D. F. Garrard (1968) The resolution of photographic emulsions for electrons in the energy range 7–60 keV. *J. Sci. Instrum.* 1:715.

——, D. F. Garrard, and M. T. Brown. (1968) The response of photographic emulsions to electrons in the energy range 7–60 keV. *J. Sci. Instrum.* 1:707.

Carroll, J. S. (1974) *Photographic Lab Handbook.* Amphoto, New York.

Eastman Kodak Co. (1973a) *Electron Microscopy and Photography*, publication number P-236. Eastman Kodak Co., Rochester, N.Y.

—— (1973b) *Gaseous-Burst Agitation in Processing*, publication number E-57. Eastman Kodak Co., Rochester, N.Y.

—— (1973c) *Processing Chemicals and Formulas*, publication number J-1. Eastman Kodak Co., Rochester, N.Y.

—— (1973d) *Kodak Filters for Scientific and Technical Uses*, publication number B-3. Eastman Kodak Co., Rochester, N.Y.

—— (1974) *Kodak Darkroom Data Guide*, publication number R-20. Eastman Kodak Co., Rochester, N.Y.

—— (1976a) *Practical Processing in B/W Photography*, publication number P-229. Eastman Kodak Co., Rochester, N.Y.

—— (1976b) *Kodak B/W Photographic Papers*, publication number G-1. Eastman Kodak Co., Rochester, N.Y.

Farnell, G. C., and R. B. Flint. (1969) Low-contrast development of electron micrographs. *J. Micros.* 89:37.

—— and R. B. Flint. (1970) A method for increasing the photographic contrast of electron micrographs. *J. Micros.* 92:145.

—— and R. B. Flint. (1973) The response of photographic materials to electrons with particular reference to electron micrography. *J. Micros* 97:271.

—— and R. B. Flint. (1975) Photographic aspects of electron microscopy. In: *Principles and Techniques of Electron Microscopy* 5:19. Hayat, M. A. (ed.). Van Nostrand Reinhold, New York.

Frieser, H., and E. Klein (1958) Properties of photographic emulsions with respect to electron radiation. *Z. Angew. Phys.* 10:337.

Guetter, E., and M. Menzel (1978) An external photographic system for electron microscopes. Proc. 9th Int. Cong. EM (Toronto), 1:92.

Hamilton, J. F., and J. C. Marchant. (1967) Image recording in electron microscopy. *J. Opt. Soc. Am.* 57:232.

—— and F. Urbach. (1966) The mechanism of the formation of the latent image. In: *The Theory of the Photographic Process*, 3rd ed. Mees, C. E. K., and T. H. James (eds.). MacMillan, New York.

Herz, R. H. (1969) *The Photographic Action of Ionizing Radiations.* John Wiley and Sons, New York.

Hillson, P. J. (1969) *Photography.* Doubleday, Garden City, N.Y.

Iwanaga, M., H. Ueyaragi, K. Hosoi, N. Iwasa, K. Oba, and K. Shiratsuchi. (1968) Energy dependence of photographic emulsion sensitivity and fluorescent-screen brightness for 100 kV–600 kV electrons. *J. Electron Micros.* 17:203.

Jacobson, C. I., and R. E. Jacobson. (1972) *Developing*, 18th ed. Focal Press, New York.

Kirkpatrick, K. (1975) *Basic Darkroom.* Peterson Publishing Company, Los Angeles.

Kumpf, U. E. (1980) Trans-fiberoptic photography speeds electron microscopy. *Ind. Res. Dev.* p. 101. July 1980.

Neblette, C. B. (ed.). (1976) *Photography: Its Materials and Processes.* Van Nostrand Reinhold, New York.

Valentine, R. C. (1966) Response of photographic materials to electrons. In: *Advances in Optical and Electron Microscopy*, vol. 1. Barer, R., and V. E. Cosslett (eds.). Academic Press, New York.

Wall, E. J., and F. I. Jordan. (1974) *Photographic Facts and Formulas.* Amphoto, New York.

3. Support Films

INTRODUCTION

Particulate specimens that are of interest to biologists include bacteria, viruses, subcellular fractions, and suspensions of macromolecules. Study of the internal morphology of these specimens requires fixation, embedding, and sectioning; study of the external topography of a specimen requires mounting of the specimen on a thin film and density enhancement for interaction with the electron beam. The next few chapters are devoted to preparing specimens for study of external morphology; Table 3-1 outlines these preparatory steps.

Table 3-1. Preparation of bacteria or viruses for study of surface morphology.

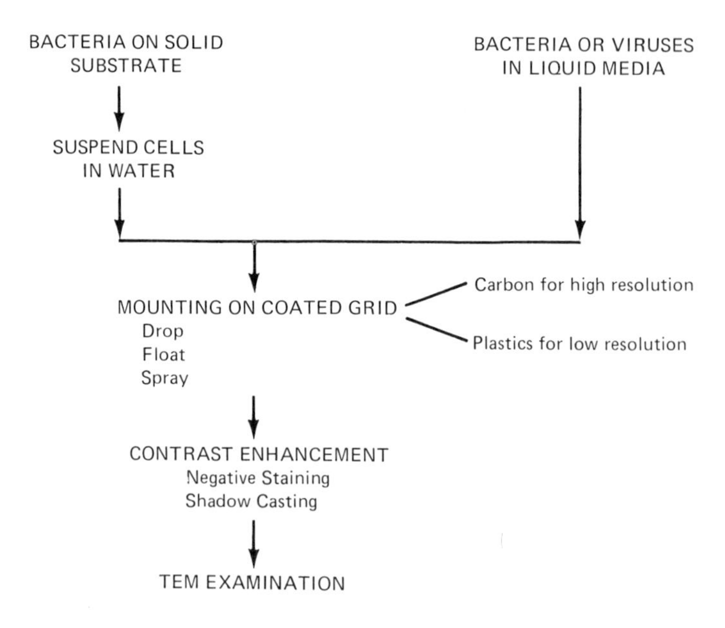

GRIDS

Electron microscope grids are analogous to glass slides for light microscopes. Any kind of sample must first be mounted on a grid before examination in a TEM, and various grid types are available for this purpose (Figure 3-1). Thin sections of tissue are placed directly on a 200- or 300-mesh grid, and serial sections are mounted on slot-type grids. Most particles, however, will fall directly through the mesh grids; consequently, such specimens are mounted on grids covered with a very thin support film.

The standard diameter of a grid is 3.05 mm, although some microscopes accept 2.30-mm-diameter grids. Copper mesh grids are most commonly used in a wide variety of applications; 200-mesh grids permit 60–70% transmission, whereas 300- or 400-mesh grids allow proportionately smaller transmissions. Some special procedures (e.g., some post-staining methods) require that grids be made from gold, silver, stainless steel, or beryllium. Athene-type grids are those that have one smooth, shiny side, while the other is matte as a result of etching. The matte side allows good adhesion of the specimen to the grid, and also reveals which side of the grid has the specimen. Thus, it is recommended that all samples be routinely mounted on the matte side of grids.

Because of their fragility, grids are initially awkward to handle. Most grids have a rim that helps in manipulation; slight bending of the edge with fine-tipped forceps is an aid. The grids are always cleaned in oil-free acetone prior to use to remove trace oil contamination and reduce static electricity. Before mounting any specimen, blot the grids dry with filter paper.

Storage units are commercially available for grids having specimens,

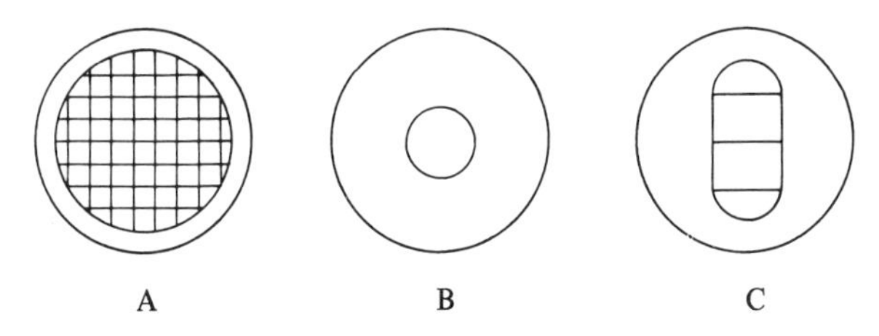

Figure 3-1. Grids for mounting of specimens. A. Mesh grid. B. Single hold grid. C. Slot grid with support.

but are expensive. One way to avoid this expense is to place a piece of filter paper into a plastic Petri dish, securing it with double-stick tape. Another piece of double-stick tape is placed to one side of the filter paper, and touching only the edge of the grid to the tape edge will secure it. The specimen may then be identified, dated, and so on, by simply writing on the filter paper. Avoid prolonged exposure of grids to the atmosphere; particles will settle on and contaminate the grid. Storage of the Petri dishes in the desiccator usually helps. Nonetheless, grids should be examined as soon as possible after preparation.

MATERIALS FOR SUPPORT FILMS

As explained above, particulate specimens must be mounted on a support film prior to examination. Support films are also used for identifying astigmatism in the microscope, and the same materials are used in preparing relicas of specimens. Only a few materials are appropriate as support films because a film must be sufficiently strong to withstand electron bombardment while still possessing a low atomic or molecular weight. Both of these criteria are important for good resolution: a strong film effects a stable sample, and a low-weight and/or ultrathin film will not scatter electrons. Various materials have been evaluated as support films (Baumeister and Hahn, 1978), with the most popular films being carbon or plastic thin films.

Plastic Thin Films

Collodion (nitrocellulose) and Formvar (polyvinyl formal) are plastics that will form support films 200–300 Å thick (Revel and Agar, 1955). Both of these are easily prepared and are used for moderate-resolution microscopy; they are not appropriate for higher-resolution work because films thinner than 250 Å will become unstable during irradiation (heat damage). Unfortunately, little work has been advanced in developing other plastics for thin films since the early 1940s. Hahn and Baumeister (1976) recommend experimentation with other radiation-resistant materials.

Collodion was introduced for electron microscopy by Ruska (1939). It is soluble in ethyl acetate, amyl acetate, or acetone in concentrations of 0–5%. Evaporation of the solvent gives rise to a polymer consisting of the following subunits:

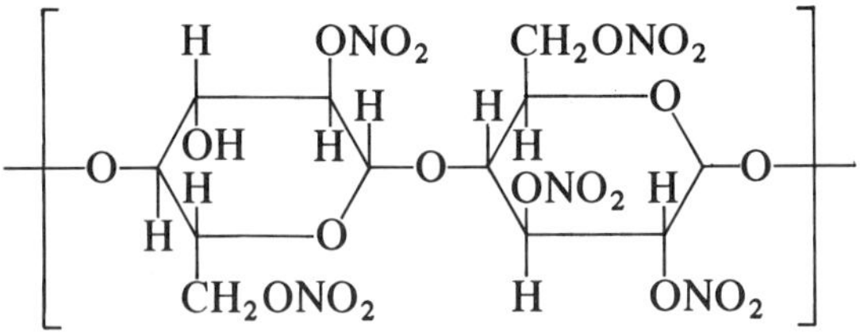

H ONO_2 CH_2ONO_2
H H O
O OH H O H O
H ONO_2 H
H O
CH_2ONO_2 H ONO_2

The advantage of Collodion is that the films exhibit little hydrophobicity, meaning that particles suspended in a liquid do not severely aggregate when that liquid evaporates away from the polymer. Collodion support films are also very smooth and may be made quite thin (~150 Å); however, films less than 300 Å thick are very unstable during irradiation and may sublime up to 85% of their mass (Reimer, 1965).

Thin Collodion films are prepared by the method of casting on water (Ruska, 1939). The basic procedure for casting on water is as follows:

1. Prepare a stock solution of 0.5–1% Collodion in ethyl acetate.
2. Arrange a funnel apparatus:
   a. Place a clean wire mesh inside the base of the funnel.
   b. Fill the funnel well above the level of the mesh with distilled water.
   c. Place clean 300-mesh copper grids, matte side up, on the mesh.
3. Place a drop of the Collodion solution in the center of the funnel above the water level, wait for the solvent to evaporate, and remove the film. This will clean the water surface.
4. Repeat step 3, but do not handle the film.
5. Slowly open the funnel stopcock, permitting water to slowly drip out of the funnel.
6. As the water level continues to drop, the film will eventually layer over the grids, evenly coating their surface.
7. Wait for the grids to dry; then carefully remove them with fine-tipped forceps.
8. Apply the sample.

Formvar, or polyvinyl formal, is more popular than Collodion because it is more radiation-resistant; only 30–40% of its mass will be lost during irradiation (Reimer, 1965). Introduced by Schaefer and Hasker (1942), it has the following structure:

$$\left[-H_2C-\underset{\substack{|\\O\\ \diagdown}}{\overset{H}{C}}-CH_2-\underset{\substack{|\\O\\ \diagup}}{\overset{H}{C}}-\right]\underset{\substack{C\\ \diagup\ \diagdown\\ H\quad R}}{} \left[-CH_2-\underset{OH}{\overset{H}{\underset{|}{\overset{|}{C}}}}-\right]\left[-CH_2-\underset{\substack{O\\|\\C=O\\|\\CH_3}}{\overset{H}{\underset{|}{\overset{|}{C}}}}-\right]$$

Formvar is soluble in concentrations of 0.1–0.5% in ethylene dichloride or chloroform, with 0.25% in ethylene dichloride a standard. Polymerized Formvar films are slightly hydrophobic, and particles tend to aggregate at the surface.

Formvar films are prepared by casting on glass; Collodion films are prepared as in the above method cast on water, or cast on glass.

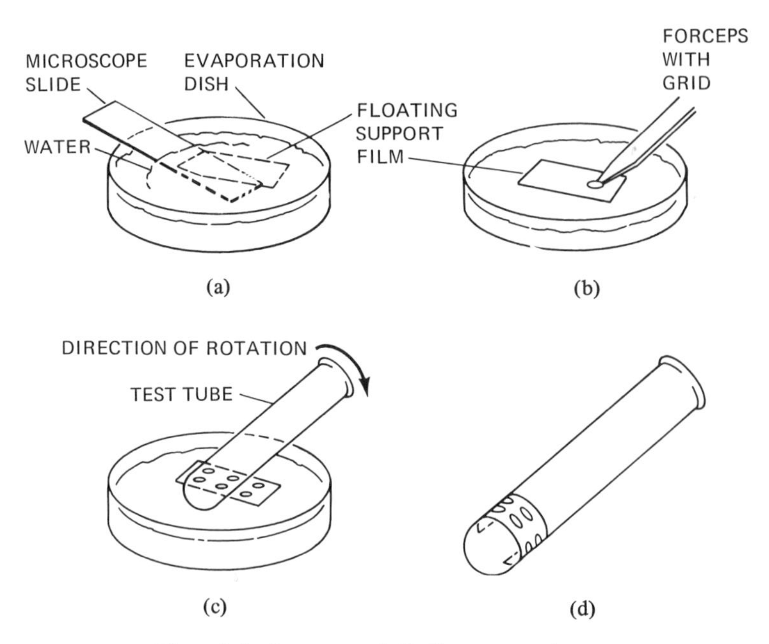

Figure 3-2. Formvar or Collodion cast on glass.

The cast-on-glass procedure is as follows (Figure 3-2) (Schaefer and Hasker, 1942):

1. Prepare a stock solution of 0.25% Formvar in ethylene dichloride.
2. Clean a glass microscope slide with soap and water, but leave a trace residue of soap, which assists in stripping the film.
3. Fill a small beaker with the Formvar stock solution, and dip the slide into the liquid. Air-dry under cover at an angle.
   a. The length of time required for drying varies with humidity; on rainy days drying is quickest if the slides are placed in a desiccator.
   b. Ethylene dichloride is volatile; lengthy exposure of the stock solution to air increases the Formvar concentration.
4. Fill an evaporation dish or fingerbowl to overflowing with distilled water, and clean the water surface by gently running a glass or Teflon rod over the surface.
5. Scrape the edges of the slide with forceps to break the film.
   a. Avoid contaminating the film surface with glass particles; scrape away from the film surface.
   b. Vigorous scraping is unnecessary; the film is only ~350 Å thick.
   c. The lower edge of the slide has a thicker film than other areas; score this edge 1–2 mm from the slide edge with a razor blade.
6. Prop the microscope slide at an angle of ~30° at the lip of the evaporation dish. (Figure 3-2a).
   a. Avoid shaking your hands by firmly placing them on the workbench.
   b. The surface tension of the water will force the thin film away from the slide. A positive meniscus greatly assists this.
7. Slowly lower the slide into the dish until the film is freely floating at the water surface. Let the slide drop into the dish. (Figure 3-2b).
   a. The film will not strip if it was wet, or if the slide was too clean.
   b. If the film is not completely released, for example, if it is stuck at a corner, tease it away from the slide with a needle.

8. Evaluate the film for contamination, wrinkles, or bright interference colors, all of which should be avoided. The best films are colorless or pale gray (see Reimer, 1967).
9. Clean, dried grids are slightly bent at the rim and placed, matte side down, on good areas of the film.
   a. Slight bending helps adhesion of the film to the grid; too much bending will distort and weaken the film.
   b. Leave at least a few millimeters between grids; do not overload the film.
10. Clean the exterior of the wide-diameter test tube, position it over the grids/thin film, push it gently into the water, rotate it, and remove it, all as one smooth motion (Figure 3-2c).
    a. Working with small films (1″ X 1″) or only at one end of a film makes this task easier.
    b. Wide-diameter test tubes avoid film overlap.
11. Dry the test tube with attached grids and film under cover (Figure 3-2d).
12. Evaluate the grids and remove them with forceps.
    a. Those grids having multiple film layers or upside-down are discarded.
    b. Slightly bent grids are easily removed by placing one tip of the forceps beneath the grid and the other tip above, and pulling the grid off at an angle perpendicular to the test tube.
13. Apply the sample to the grid.

Although some practice may be necessary to successfully prepare Formvar-coated grids, once the technique is mastered it is very rapidly performed. This method may be modified to produce "holey films" for diagnosis of astigmatism in the microscope (Harris, 1962). Introducing small amounts of water into the liquid Formvar will cause many holes of various diameter in the film because water droplets interfere with polymerization. Baumeister and Hahn (1978) discuss various preparations, but the simplest way requiring only a small volume of material is as follows: in step 3 of the plastic-cast-on-glass method, immediately after removing the slide from the stock solution, breathe on the slide. Microdroplets from this warm breath deposit on the slide and form small holes. Continue the preparation as given. Shadowing the holey film, which will be discussed, stabilizes the film and the grid can then be repeatedly used for diagnosing and correcting astigmatism.

Both Formvar and Collodion stock solutions rapidly degenerate; this deterioration can be slowed by storing the solutions in tightly-capped amber bottles. Contaminated solutions exhibit any or all of the following: cloudiness, very thick films showing only bright interference colors, severe problems with stripping, or holes in the film. Coated grids should be prepared just prior to use, although if absolutely necessary they may be stored up to ~30 days. Care should be taken to prevent contamination by settled particles.

Because of their instability, a few precautions should be observed in examining plastic films. Low beam current and intensity lessen tearing of film, and a given field-of-view should not be exposed for extended periods of time. Photography should be rapid; if the specimen is drifting during the prolonged film exposure, the negative will be out of focus. Finally, be patient when using plastic films because their advantages outweigh their disadvantages for routine lab work.

Carbon Support Films

A carbon film is the best type of support for high-resolution microscopy. By evaporating carbon from an arc under vacuum, very thin, strong films are obtained. The films are amorphous; that is, they do not possess a regular crystalline pattern (Kakinoki et al., 1960). Another feature of carbon is its low atomic weight, meaning that it does not interact with the electron signal. Finally, carbon films are extremely strong while also very thin: a 100-Å-thick film is virtually unbreakable in the TEM (accelerating voltage ~125 keV). Film thickness may be varied from 20 to 100 Å and still exhibit uniformity (Moretz et al., 1968).

Bradley (1954) introduced evaporated carbon films for electron microscopy. The apparatus used for evaporation is the vacuum bell jar, which is essentially a large glass jar evacuated by a rotary and diffusion pump (Figure 3-3). Ports for passage of high-voltage cables are available through the baseplate of the bell jar. Spectroscopically pure carbon electrodes are held end to end within brass holders: one electrode is spring-loaded, while the other is immobile. The tension

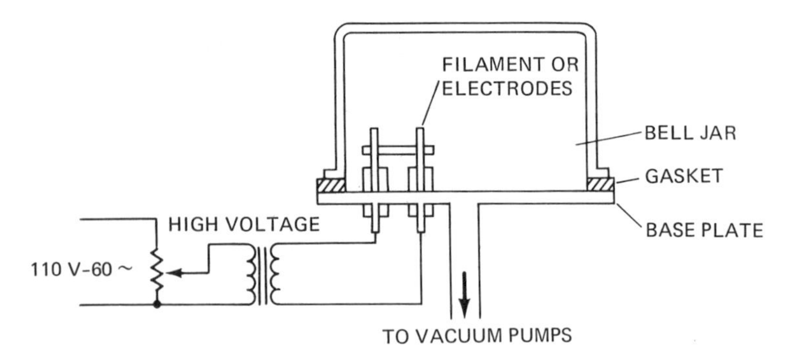

Figure 3-3. Vacuum bell jar apparatus.

is necessary to maintain a constant contact between the two electrodes during evaporation; the degree of tension controls film thickness. Different electrode configurations are used, although apparently there is no specific advantage of one over another. The arrangement shown in Figure 3-3 is of two pointed electrodes, which work well but are sometimes difficult to align exactly without damage. An arrangement easier to work with has one pointed end held against a squared-off end; the pointed electrode is spring-held.

The target on which the thin film is to be deposited is located 10–15 cm beneath the electrodes; as will be discussed, freshly cleaved mica is used as the substrate. Near the target is a clean white porcelain plate with a drop of vacuum oil or grease on it, to assist in estimating film thickness (Bradley, 1954); the area covered by the drop will remain white, while the surrounding area will show shades of gray, with dark gray being thick. After evacuation to $\sim 10^{-4}$ Torr, an alternating current of 30 amp at 15 V is passed through the electrodes. This is sufficient to heat the electrodes to white hot, and the thermal energy generated forces carbon atoms off the electrodes. The atoms will travel in straight lines until striking the interior of the bell jar, where, because of its much lower temperature, the carbon will be trapped. Thus, the target will be gradually covered with individual atoms, and over time (~10–30 sec) a continuous thin film will be formed. (Further information on the growth of thin films will be presented in Chapter 6.)

DeBoer and Brakenhoff (1974) have evaluated film thickness in a more careful manner than the above method, by measuring the degree of light absorption as a function of the amount of carbon evaporated. Using their method, one may reproduce films predictably.

Variations of the above-described method have improved the quality of carbon films in terms of graininess and smoothness. By applying current in short pulses rather than continuously, the deposition rate of the atoms is better controlled (Baumeister and Hahn, 1978). A more elaborate scheme involves deflecting the carbon atoms before they strike the target, thus ensuring that single atoms rather than clusters contribute to thin film formation (Johansen, 1974).

Some different targets have been used as substrates for carbon films, namely, glass microscope slides, freshly cleaved sheets of mica, and grids with plastic support films. A good target should be smooth and structureless. A smooth substrate ensures easy stripping of the film, a problem when using glass slides: commercially available pre-cleaned slides frequently have a residual layer of detergent that interferes with removal, and slides may also be inherently rough-surfaced. Bradley (1967) recommended detergent as an aid to stripping, but it is not recommended here because it is a likely source of contamination. About the only advantage of evaporation onto glass slides is that they are readily available.

A substrate that avoids these problems is a freshly cleaved sheet of mica (Hall, 1956; Spencer, 1959; Williams and Glaeser, 1972). Sheets of mica are commercially available from EM suppliers, and virtually the only treatment necessary before evaporation is to cleave the mica into small squares (~1″ × 1″). This is done by gently introducing a razor edge or forceps tip between two crystal planes, and exerting slight pressure until the natural cleavage planes separate. Take care not to force the crystals apart or scratch them; gentle, continuous pressure will result in two very smooth targets. It is also recommended that the crystals not be paper-thin; a slightly thicker target is easier to handle. Mica targets are prepared just prior to use.

Removal of a thin film from a mica square or glass slide is similar to casting on glass. The procedure is:

1. Arrange clean grids and an evaporation dish with distilled water as in the cast-on-glass method.
2. Grasp the mica/glass and lower it into the dish at a shallow angle, leaving the film floating on the water surface. Mica

targets are easily handled if locking forceps are used for manipulation.

3. Lower a clean 200- or 300-mesh grid below the water surface, position it beneath and parallel to the film, and gently pull it directly upward through the film.
4. While the grid/film is still wet, blot the backside on filter paper to remove carbon that has wrapped around it.
5. Air-dry under cover; grids may be stored for a short period of time in a Petri dish with slightly damp filter paper.

Carbon films may be evaporated directly onto plastic-coated grids (Bradley, 1967); the plastic is subsequently dissolved, and the carbon adheres to the grid. The only advantage of this method is that residual plastic on the grid bars glues the carbon to the grid; however, its disadvantages are overwhelming. First, the carbon film replicates the surface features of the plastic film (i.e., any imperfection will be reproduced). Second, dissolution of the plastic requires around 12 hr or longer of treatment, because the carbon–plastic–grid sandwich is exposed only to solvent vapors. The normal procedure is to saturate a few sheets of filter paper with the appropriate solvent in a Petri dish, place a metal mesh over the paper, and put the grid on top of this. Dissolution is complete after overnight exposure.

Carbon films are prepared immediately prior to use because they inherently are hydrophobic, and this property will be exaggerated over time. Thus particles will tend to clump on carbon films. The next chapter discusses methods of applying a particulate suspension to a grid, and those that are least affected by hydrophobicity should be used (i.e., the float or spray method; Dubochet and Kellenberger, 1972). One way to lessen the hydrophobic nature is to place carbon-coated grids on moist filter paper in a Petri dish between their time of preparation and their use.

Baumeister and Hahn (1978) provide a comprehensive review of support films, including those used for supporting single atoms, such as aluminum oxide (Koller, 1971; Muller and Koller, 1972) and graphite films (Fernandez-Moran, 1960; Beer and Highton, 1962; Dobelle and Beer, 1968; Hashimoto et al., 1974; Millward et al., 1978; Crewe et al., 1980). Their review also discusses materials that were tried but rejected as potential supports, for example, glass

(Hart, 1966), mica (Heinemann, 1970), and beryllium (Cosslett, 1948; Komoda et al., 1969).

REFERENCES

Baumeister, W., and M. Hahn. (1978) Specimen supports. In: *Principles and Techniques of Electron Microscopy* 8:1. Hayat, M. A. (ed.). Van Nostrand Reinhold, New York.

Beer, M., and P. J. Highton. (1962) A simple preparation of graphite-coated grids for high resolution electron microscopy. *J. Cell Biol.* 14:499.

Bradley, D. E. (1954) Evaporated carbon films for use in electron microscopy. *Br. J. Appl. Phys.* 5:65.

—— (1967) The preparation of specimen support films. In: *Techniques for Electron Microscopy*, 2nd ed., p. 58. Kay, D. H. (ed.). Blackwell Scientific Pub., Oxford.

Cosslett, V. E. (1948) Beryllium films as object supports in the electron microscopy of biological specimens. *Biochem. Biophys. Acta* 2:239.

Crewe, A. V., R. K. Mittleman, and R. L. Rosman. (1980) Preparation of thin graphite supporting films. *Proc. 38th Ann. EMSA Meet.*, p. 410.

De Boer, J., and G. J. Brakenhoff. (1974) A simple method for carbon film thickness determination. *J. Ultrastr. Res.* 49:224.

Dobelle, W. H., and M. Beer. (1968) Chemically cleaved graphite support films for electron microscopy. *J. Appl. Phys.* 35:1652.

Dubochet, J., and E. Kellenberger. (1972) Selective adsoption to the supporting film and its consequences on particle counts in electron microscopy. *Micros. Acta* 72:119.

Fernandez-Moran, H. (1960) Single crystals of graphite and of mica as specimen supports for electron microscopy. *J. Appl. Phys.* 31:1840.

Hahn, M., and W. Baumeister. (1976) New specimen supports for high and ultra high resolution. *Developments in Electron Microscopy and Analysis.* Academic Press, London.

Hall, C. E. (1956) Visualization of individual macromolecules with the electron microscope. *Proc. Nat. Acad. Sci.* 42:801.

Harris, W. J. (1962) Holey films for electron microscopy. *Nature* 196:499.

Hart, R. G. (1966) Glass supporting film for electron microscopy. *J. Appl. Phys.* 37:3315.

Hashimoto, H., A. Kumao, H. Endon, and A. Ono. (1974) Image contrast of atoms and substrate films in bright and dark filed technique. *Proc. 8th Int. Cong. EM* (*Carberra*) I:244.

Heinemann, K. (1970) A comment on mica as electron microscope support film. *28th Ann. EMSA Proc.*, p. 526.

Johansen, B. V. (1974) Bright field electron microscopy of biological specimens. II. Preparation of ultra-thin carbon support films. *Micron* 5:209.

Kakinoki, J., K. Katada, T. Hanawa, and T. Ino. (1960) Electron diffraction study of evaporated carbon films. *Acta Cryst.* 13:171.

Koller, T. (1971) Suitability of aluminum oxide support films for high resolution electron microscopy. *Proc. 15th Ann. Meet. Biophys. Soc. (New Orleans)*: *Biophys. J.* 11:216a.

Komoda, T., J. Nishida, and K. Kimota. (1969) Beryllium single crystal flakes as substrates for high resolution electron microscopy. *Jap. J. Appl. Phys.* 8:1164.

Millward, G. R., D. A. Jefferson, and J. M. Thomas. (1978) On the feasibility of imaging, by high resolution electron microscopy, isolated individual sheets of graphite carbon. *J. Micros.* 113:1.

Moretz, R. C., H. M. Johnson, and D. F. Parsons. (1968) Thickness estimation of carbon films by electron microscopy of transverse sections and optical density measurements. *J. Appl. Phys.* 39:5421.

Muller, M., and T. Koller. (1972) Preparation of aluminum oxide films for high resolution electron microscopy. *Optik* 35:287.

Reimer, L. (1965) Irradiation changes in organic and inorganic objects. *Lab. Invest.* 14:1082.

—— (1967) *Electronenmikroskopische Untersuchungs und Präparationsmethoder*, 2. Aufl. Springer-Verlag, Berlin, New York.

Revel, R. S. M., and A. W. Agar. (1955) The preparation of uniform plastic films. *Br. J. Appl. Phys.* 6:23.

Ruska, H. (1939) Ubermikroskopische Untersuchungstechnit. *Naturwissenschaften* 27:287.

Schaefer, V. J., and D. Hasker. (1942) Surface replicas for use in the electron microscope. *J. Appl. Phys.* 13:427.

Spencer, M. (1959) The preparation of carbon support films for electron microscopy. *J. Biophys. Biochem. Cytol.* 6:125.

Williams, R. C., and R. M. Glaeser. (1972) Ultra thin carbon support films for electron microscopy. *Science* 175:1000.

4. Particulate Specimen Application

After preparing grids with support films, it is necessary to mount particulate specimens in such a manner that the sampling is representative of the bulk specimen and the particles do not overlap or in any way interfere with one another. The methods used for mounting wet suspensions are the drop, float, and spray methods. As discussed in Chapter 5, these same methods are used for application of a negative stain for contrast enhancement.

The choice of a specific method of particle application is partially dependent on the concentration of the suspension. In general, a concentration of 0.1 mg/1 ml is more than sufficient (Haschemeyer, 1968, 1970; Haschemeyer and Meyers, 1972) when handling macromolecular suspensions. Horne (1967c) determined that dilution of buffered enzymes 20 to several hundred times with distilled water may be tolerated, using an initial concentration of 0.5–5 mg/ml, provided that the enzyme does not dissociate with a change in pH. In comparison, macromolecules and viruses that are sensitive to changes in pH require buffer dilutions—a problem because many buffers tend to precipitate during drying and interfere with specimen observation. One method for avoiding precipitates is using phosphotungstic acid as the negative stain; it tolerates, for example, phosphate buffers (Horne, 1967b). Horne also recommends that several dilutions (e.g., 50-, 100-, 200-, and 500-fold) be independently prepared for evaluation of the optimum.

Surface studies of bacteria are easy to conduct. Bacteria grown on a solid substrate (e.g., agar slants or solid cultures) are suspended in water by removing a small portion of the colony with an inoculating loop and stirring this in a few milliliters of distilled water. A concentrated suspension results, and grids should be prepared by the float technique (below).

Bacteria growing in liquid cultures present the same problems as buffered viruses or macromolecules. To separate the cells from the detritus, centrifugation several times with resuspension and intermediate culture-medium washes or distilled water washes (if tolerable to bacteria) yields a pure suspension of cells. The suspension is then diluted as above, and grids are prepared by dropping, floating, or spraying. This method is well suited for cells to be shadowed, where it is essential that no artifacts be present.

However, with a great deal of luck, it is sometimes possible to use the liquid culture as the actual suspension to be applied to a grid. Fresh cultures do not typically have excess matter present; grids are floated on a drop of the culture and negatively stained, again by the float method. As will be discussed, this method is very rapidly performed, and grids may be immediately examined and evaluated. If too much detritus is visible, cells should be purified as above.

The drop method simply refers to placing a drop of the dilute specimen suspension on a support film/grid. Application is with a small pipette, or platinum loop; the drop remains on the grid for approximately 30 sec (standard), or longer if dilute suspensions are used (Haschemeyer and Meyers, 1972). The excess liquid is withdrawn with a torn piece of filter paper, and the grid either immediately negatively stained and then dried, or dried for shadow casting.

A variation of the drop method is the float technique (Haschemeyer and Meyers, 1972). It simply involves placing a drop of the sample suspension in a plastic Petri dish (or on a piece of Parafilm) and floating the supported grid for 30 sec (standard) or longer. Particles will be attracted by electrostatic forces and adhere to the support film. This method is best used when concentrated samples are available because if the grid is then floated on a drop of negative stain, a fraction of the particles will diffuse into the stain and be lost. Use of this method typically offsets the problem of excess sample.

Two major problems may be encountered in using the drop and float method. First, as the specimen is going through the final stages of air drying, it will undergo tremendous stress from interfacial tension. Air drying causes distortion from volume and/or surface changes. In relatively large cells (e.g. bacteria), the volume-to-surface ratio is large, and volume stresses are dominant. Consequently, these large cells tend to collapse or deform. Smaller specimens (e.g., viruses and macromolecules) have a proportionately smaller volume-to-surface ratio, and are subject to surface forces. Interfacial tension is the force set up by a receding liquid meniscus; Anderson (1951) has calculated the stress in air-drying a bacterial flagellum to be an incredible 46,000 kg/cm^2.

Negative staining will alleviate but not eliminate the interfacial tension, primarily because it will strengthen the specimen. When the negative stain is applied to the grid, it will fill in any irregularities on the surface of the specimen as well as the area between the specimen and support film. As the stain solvent evaporates, the heavy metal phase of the stain solidifies around the particle, thus molding and stabilizing it. For this mechanism to work, the negative stain must be applied to a wet specimen grid; if the grid dries between specimen application and staining, the damaged particles will be preserved.

Other problems, which are more frequently encountered with the drop than the float method, revolve around the distribution of the particles on a grid. Particles tend to aggregate in concentrated samples, but, conversely, low concentrations result in few observable particles. Ideally, discrete particles are desired. Recall that the majority of a grid is closed space; thus the optimum concentration for a given specimen is best obtained through trial and error.

Some of these problems may be reduced by employing the spray technique, which involves the deposition of microdroplets by a nebulizer (Backus and Williams, 1949). Figure 4-1 is a simple Vaponefrin nebulizer, consisting of a large glass bulb within which there is a diffusor head. A small volume of sample (0.1–0.2 ml) is placed into the glass chamber via a cork-sealed port, grids with support films are positioned 3–10 cm from the exit of the outlet tube, and a stream of microdroplets is initiated by squeezing the rubber bulb. Microdroplets 5–20 μm in diameter will be deposited on the support film/grid.

An advantage of the spray technique is that because the liquid volume per microdroplet is very small, interfacial forces are less than

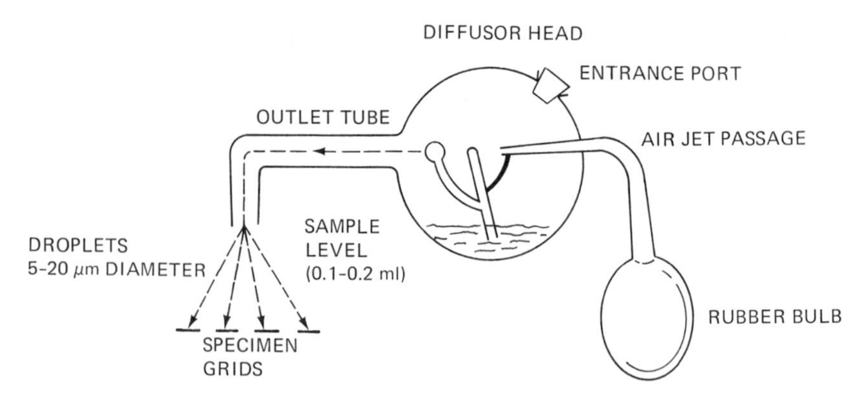

Figure 4-1. Vaponefrin glass nebulizer.

in the drop or float technique. A very good particle distribution also results; this may be enhanced by spraying a mixture of the specimen and negative stain simultaneously on the grid. Finally, only very small volumes of sample are necessary. Horne (1967a,b,c) provides an excellent review of other spray methods (e.g., those dealing with quantitative particle counting and handling infectious materials).

Regardless of the method used to apply the sample, a few other factors will influence success or failure. A major parameter is the degree of hydrophobicity possessed by the support film. Recall that carbon is the most hydrophobic, Formvar is intermediate, and Collodion is the least hydrophobic film. In all films the degree of hydrophobicity increases over time; consequently, one should prepare support films, apply the sample, and enhance contrast consecutively without significant periods of time between stages.

Another parameter, which should be tested for each type of specimen, is whether or not different treatments will enhance results. For example, will shortening or lengthening the amount of time the grid is in contact with the specimen improve the distribution patterns? This can only be evaluated by the individual researcher, usually by trial-and-error methods.

REFERENCES

Anderson, T. F. (1951) Techniques for the preservation of 3-dimensional structures in preparing specimens for the electron microscope. *Trans. N. Y. Acad. Sci.* 13:130, 132.

Backus, R. C., and R. C. Williams. (1949) The use of spraying methods and of volatile suspending media in the preparation of specimens for electron microscopy. *J. Appl. Phys.* 21:11.

Haschemeyer, R. H. (1968) Electron microscopy of enzymes. *Trans. N. Y. Acad. Sci.* 30:875.

—— (1970) Electron microscopy of enzymes. In: *Advances in Enzymology* 33:71. Nord, P. F. (ed.), John Wiley and Sons, New York.

——, and R. J. Meyers, (1972) Negative staining. In: *Principles and Techniques of Electron Microscopy: Biological Applications,* vol. 2. Hayat, M. A. (ed.), Van Nostrand Reinhold, New York.

Horne, R. W. (1967a) The examination of small particles. In: *Techniques for Electron Microscopy,* 2nd ed. Kay, D. H. (ed.), Blackwell Scientific Pub., Oxford.

—— (1967b) Negative staining methods. Ibid.

—— (1967c) Electron microscopy of isolated virus particles and their components. In: *Methods in Virology.* Maracorosch, K., and H. Koprowski (eds.), Academic Press, New York.

5. Negative Staining

INTRODUCTION

The modern electron microscope is capable of a point-to-point resolution of only a few angstroms, as demonstrated by micrographs of gold or graphite crystals. Thus, the resolution of the instrument does not typically limit the study of biological ultrastructure; the sample preparation methods are the parameters controlling resolution (Horne, 1973). Biologists are consequently challenged with preparing samples that are faithful to life, protecting them against vacuum exposure and radiation damage, and finally introducing sufficient density to the specimens for electron interaction.

A prerequisite for the study of the surface morphology of particles is an increase in density, resulting in image contrast (Horne, 1967; Haschemeyer, 1968). After the specimen (viruses, bacteria, or macromolecules) is mounted on a coated grid, density is enhanced by either negative staining with heavy metals in solution, or shadow casting of high-molecular-weight metals by evaporation. These techniques ensure that sufficient density is imparted to the specimen for interaction (scattering) with the electron beam.

THEORY

Negative staining increases the density of the area surrounding discrete particles, as opposed to positive staining, which is a reaction between the stain and specimen (Figure 5-1). Therefore, a negatively stained specimen will be viewed in negative contrast. A proper understanding of the mechanism of negative staining involves first a discus-

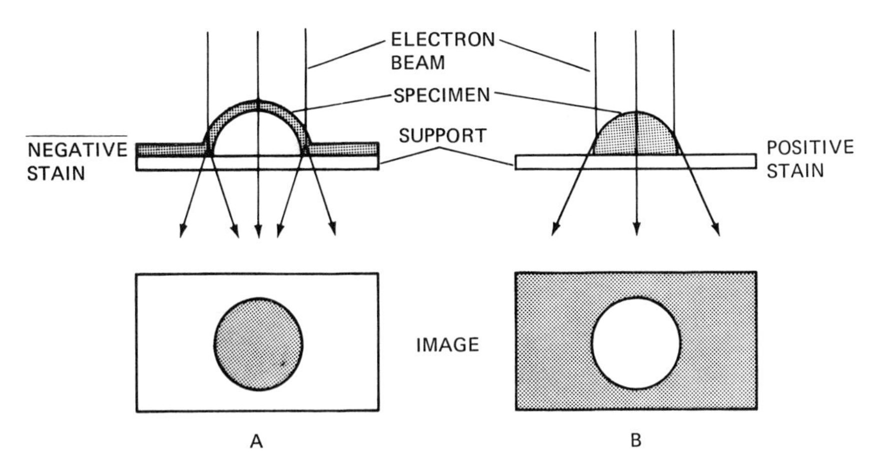

Figure 5-1. A comparison of negative staining (A) and positive staining (B).

sion of how any heavy metal stain enhances contrast. The major goal of staining is to separate the electron-opaque from electron-transparent regions, (i.e., pull the specimen out from the background). Unstained biological materials are inherently poor electron scatterers, because of their low-atomic-weight constituents. Valentine and Horne (1962) showed that an object is resolvable only if the product of thickness (Å) and weight density (g/cm^3) exceeds 400; the object may be resolved if the product is as low as 100, but only poorly. They assume a density of approximately 1 for most biological materials, meaning that at best, without contrast enhancement, a resolution of only 100 Å can be obtained at optimal operation.

Enhancement of contrast by positive staining will increase the resolution, but a give-and-take situation rapidly develops. Doubling the weight density will increase resolution to about 50 Å, but serious distortion of the specimen results. On the other hand, distortion is eliminated if the background density of the specimen is increased. When a heavy metal solution three times greater in density is applied to a biological sample of density 1, the background will be three times greater than its natural density. As a result of this negative reaction, particles only 33 Å in diameter should be visible (Valentine and Horne, 1962).

When a solution of a negative stain is applied to a grid, the stain coats the specimen, reproducing its topography and filling in the background. As the solvent evaporates, it leaves behind a smooth film of heavy metal. Because the specimen is held within this dense supporting matrix, the specimen is preserved (Johnson and Horne, 1970).

Negative staining has been employed in light microscopy for nearly a century; Hall (1955) was the first to employ negative staining in the electron microscopy of viruses. Subsequently, negative stains have found diverse applications in the study of enzymes (Haschemeyer, 1968), proteins (Mellema et al., 1967a,b; Gordon et al., 1974; Malech and Albert, 1979), cellular fractions (Muscatello and Horne, 1968; Muscatello et al., 1972a,b, 1975; Racker, 1978), bacteria (Haschemeyer and Meyers, 1972), and viruses (Huxley, 1956; Horne and Pasquali-Ronchetti, 1974; Horne et al., 1975). A number of specific methods have been summarized by Horne (1967) for preparing samples by negative staining.

The unique advantage of negative staining is that it is very rapidly and simply performed; rough sample preparation time is a few minutes. Unfortunately, it is unpredictable and requires low particle concentrations (around 0.1 mg/ml). However, when negative staining is successful, it works extremely well.

NEGATIVE STAINS

Haschemeyer and Meyers (1972), in their comprehensive review of negative staining, provided the properties of a good negative stain, as did Pease (1964). First and most important, a negative stain does not react with the sample. Whereas one stain may be perfectly suited for a given sample, it may be absolutely unusable for another. Bacteria possessing slime layers or capsules pose special problems in that these extraneous coats may be distorted by reaction with the stain. This property defines the success or failure of negative staining for a given type of specimen.

The stain must be of sufficient density to evoke good electron scatter, and simultaneously its molecules should be small enough to permit high resolution of surface ultrastructure. It should also possess a melting point well above the temperature induced by electron beam heating.

These apparent contradictions are compounded by the final characteristic of a good negative stain, high solubility in water. Although only low concentrations (0.5–2.5%) of negative stain are placed on the specimen, during drying as the solvent evaporates the heavy metal salt concentration will proportionately increase. An amorphous, smooth film of stain will result only if the stain does not crystallize until the last possible moment of evaporation. Valentine and Horne (1962) determined that solubility should be at least 80 g/100 ml to fulfill this requirement.

The most popular negative stains used in TEM are phosphotungstic acid, uranyl acetate, uranyl oxalate, ammonium molybdate, and sodium silicotungstic acid (see Table 5-1). These are salts of heavy metals that fulfill the above requirements.

Phosphotungstic acid (abbreviated KPT or PTA) is widely used as a negative stain in both light and electron microscopy. Quintarelli et al. (1971) summarize very well the characteristics and applications of PTA. It has been used as a positive stain for tissue proteins (Silverman and Glick, 1969), various tissues (Huxley, 1959), and cell surfaces (Dermer, 1973). Other researchers have employed PTA as a negative stain for specimens such as viruses (Watson, 1962a,b; Watson et al., 1963). The unique advantage of PTA is that it can be pH-adjusted within a pH range of 6–8. As with any negative stain, the pH should be adjusted to that of the medium in which the specimen is suspended. Another advantage is that PTA can withstand phosphate buffers, and thus may be used when phosphate-based broth suspensions are employed. It may also be prepared in the concentration range 0.5–5.2%, although around 2% is most commonly used. The shelf life of stock solutions is quite good, although a slight drop in pH occurs. Thus, before use, always check the pH of the PTA and adjust it to the desired level.

The negative staining of proteins is best performed using uranyl salts. Uranyl acetate stains best below the isoelectric point (Van-Bruggen et al., 1960) and uranyl oxalate at or above the isoelectric point (Mellama et al., 1967b). Koller et al. (1973) thoroughly discuss uranyl staining of nucleic acids, while Hayat (1975) covers the chemistry of uranyl staining. The very short shelf life of either of these solutions may be extended by freezing, in the dark, small vials of the stock solution.

Uranyl acetate is water-soluble up to a concentration of about 1%

Table 5-1. Characteristics of some negative stains.

NAME	MOLECULAR FORMULA	MOLECULAR WEIGHT	CONC. RANGE (%)	PH RANGE	USES
Ammonium molybdate	$(NH_4)_2MoO_4$	196.01	1–3	6–8	general; microsomes
Phosphotungstic acid (KPT, PTA)	$H_3[P(W_3O_{10})_4] \cdot 14H_2O$	6498.93	0.5–5.2	6–8	general; nucleus
Sodium silicotungstic acid (STA)	$Na_4[Si(W_3O_{10})_4] \cdot 2OH_2O$	3326.53	1–4	6.8–7.2	general
Uranyl acetate	$UO_2C_2O_4 \cdot 3H_2O$	412.09	0.5–1	4–5.7	excellent for staining below isoelectric point

at a pH range of 4–5.7 (Watson, 1958). It results in very high contrast and simultaneously does not introduce any of its own structural features. More will be said about uranyl acetate in upcoming chapters; it is also used as a fixative, positive en bloc stain, and post-stain for tissues (Silva et al., 1968; Terzakis, 1968).

Uranyl oxalate negatively stains proteins and viruses above the isoelectric point (Mellama et al., 1967a,b) at the pH range 6.5–6.8. Apparently because of its lower molecular weight when compared to PTA, uranyl oxalate readily fills in minute spaces, leading to good resolution.

Sodium silicotungstic acid and ammonium molybdate are similar to PTA in terms of staining properties. (Valentine, 1961; Valentine and Horne, 1962; Haschemeyer, 1970). Sodium silicotungstic acid has been used for negative staining of various macromolecules, especially proteins (Valentine and Horne, 1962). Its concentration range is 1–4% at pH 6.8–7.2. An advantage of this negative stain is that it contains the same molar fraction of tungsten as does PTA, making it one of the highest-molecular-weight stains available.

Ammonium molybdate (concentration range 1–3%, pH range 6–8) negatively stains cell fractions very well; Muscatello and Horne (1968) have used it for enhancing the contrast of red blood cell membranes, microsomes, and mitochondria. The advantage of ammonium molybdate is that tonicity may be adjusted; a 1% solution of ammonium molybdate has tonicity comparable to 0.12 M sucrose (Muscatello and Horne, 1968). It has also been used for negative staining of catalase crystals, which are used to calibrate magnification of the TEM (Haschemeyer and Meyers, 1972).

METHODS OF NEGATIVE STAINING

A number of methods have been developed for applying a negative stain to a specimen, all of which basically involve a drop, float, or spray method. Either the suspended sample is mixed with the stain and then transferred to a grid, or a sample is mounted on a coated grid and the stain subsequently applied.

The drop, float, and spray methods are identical to those discussed in Chapter 4. During these methods, it is extremely important that the specimen grid remain wet between the time of mounting the sample and the staining (Gregory and Pirie, 1973). If the grid is allowed to dry between these steps, nonuniform negative staining will result because the stain solution cannot overcome surface tension forces dominant on the grid (recall the problem of hydrophobicity of carbon grids).

The basic procedure for the float technique is as follows:

1. Prepare carbon-, Formvar-, or Collodion-coated grids.
2. In a plastic Petri dish, place a drop of the sample suspension, and within the same dish but separate, place a drop of the desired negative stain.
3. Float the grid, support side down, on the drop of specimen for about 30 sec (variable as a function of specimen concentration).
4. Remove the grid and rapidly blot the excess suspension by touching the grid edge to a piece of ashless filter paper.
5. Immediately float the grid on the drop of negative stain for 30 sec (standard).
6. Repeat steps 2 through 5 on different grids to ensure that an optimum staining duration is reached (as long as 10 min. may be necessary).
7. Dry the grids under cover and examine them.

The float method is best used when concentrated (about 0.1 mg/ml) samples are available. Electrostatic forces will attract the specimen to the grid and hold them; some particles will be dislodged during the staining step, but typically good distributions result.

Sample loss, important when low concentrations of specimen are available, is prevented by using either the drop or the spray method. The drop technique is simply a variation of the float method, as follows:

1. Prepare carbon or plastic-coated grids. When carbon films are used, take care not to physically damage the grid during handling.
2. Using a micropipette or platinum loop, place a drop of the sample suspension on the grid for 30 sec (very dilute specimens require longer times).
3. Blot off the excess sample, and immediately apply a drop of the desired stain for 30 sec (again may be increased).

The spray method applies when only very small volumes of specimen are available. In general, most researchers have not observed any great differences between this rather elaborate method and the simpler methods (Haschemeyer and Meyers, 1972); nonetheless, some have produced quite good results when a mixture of sample/negative stain is sprayed on a grid (Brenner and Horne, 1956). Horne (1967) has discussed the various spray procedures in detail.

Mixing the stain with the sample and applying this mixture to a grid is useful when highly unstable particles are to be studied. For example, some proteins rapidly dissociate unless pH is carefully monitored; mixing the solution with the negative stain affords some stability. Another case in which a mixture is useful is that of concentrated specimens; dilution and staining may be simultaneously performed.

The evaluation of negatively stained specimens is highly subjective, with aesthetic qualities typically distinguishing a good from a bad preparation. The major reason for this is that negative staining does not always work. Apparently the degree of "wetness" of the grid during transfer to the stain greatly influences results, as will staining duration. Some grids will exhibit an excellent stain pattern in one area, but be over- or understained in another. This is the major reason why multiple runs for a given specimen should be performed. A well-stained specimen shows discrete particles that are not collapsed or distorted in any way; frequently a halo of stain surrounds the particle and continues beyond it in a pattern.

Finally, a good preparation does not show artifacts of electron-dense precipitates. It is essential that aseptic techniques be employed throughout preparation; artifacts will detract from an otherwise acceptable micrograph. Careful monitoring of stain pH, age, etc., use of purified sample suspensions, and examination of the grids as soon as possible eliminate many of these problems.

See the Appendix for preparation of negative stains and staining methods.

REFERENCES

Brenner, S., and R. W. Horne. (1956) A negative staining method for high resolution electron microscopy of viruses. *Biochem. Biophys. Acta* 34:103.

Dermer, G. B. (1973) Specificity of phosphotungstic acid used as a section stain to visualize surface coats of cells. *J. Ultrastr. Res.* 45:183.

Gordon, C. N., M. Skikite, and P. F. Hall. (1974) The use of a novel crosslinking procedure in demonstrating the subunit structure of an oligomeric protein by negative staining. *J. Ultrastr. Res.* 47:285.

Gregory, D. W., and B. J. S. Pirie. (1973) Wetting agents for biological electron microscopy. I. General considerations and negative staining. *J. Micros.* 99:251.

Haschemeyer, R. H. (1968) Electron microscopy of enzymes. *Trans. N. Y. Acad. Sci.* 30:875.

—— (1970) Electron microscopy of enzymes. In: *Advances in Enzymology* 33:71. Nord, P. F. (ed.). John Wiley and Sons, New York.

—— and R. J. Meyers. (1972) Negative staining. In: *Principles and Techniques of Electron Microscopy* 2:101. Hayat, M. A. (ed.). Van Nostrand Reinhold, New York.

Hall, C. E. (1955) Electron densitometry of stained virus particles. *J. Biophys. Biochem. Cytol.* 1:1.

Hayat, M. A. (1975) *Positive Staining for Electron Microscopy.* Van Nostrand Reinhold, New York.

Horne, R. W. (1967) Negative staining methods. In: *Techniques for Electron Microscopy,* 2nd ed., p. 328. Kay, D. H. (ed.). Blackwell Scientific Pub., Oxford.

—— (1973) Contrast and resolution from biological objects examined in the electron microscope with particular reference to negatively stained specimens. *J. Micros.* 98:286.

—— and I. Pasquali-Ronchetti. (1974) A negative staining-carbon film technique for studying viruses in the electron microscope. I. Preparative procedures for examining icosahedral and filamentous viruses. *J. Ultrastr. Res.* 47:361.

——, I. Pasquali-Ronchetti, and J. M. Hobart. (1975) A negative staining-carbon film technique for studying viruses in the electron microscope. II. Application to adenovirus type 5. *J. Ultrastr. Res.* 51:233.

Huxley, H. E. (1956) Some observations on the structure of tobacco mosaic virus. *Proc. Europ. Conf. EM,* p. 280.

(1959) Some aspects of staining of tissue for sectioning. *J. Roy. Micros. Soc.* 78:30.

—— and G. Zubay. (1960) Fixation and staining of nucleic acids for electron microscopy. *Proc. Eur. Reg. Conf. EM,* p. 699.

Johnson, M. W., and R. W. Horne. (1970) Some observations on the relative dehydration rates of negative stains and biological objects. *J. Micros.* 91:192.

Koller, T., M. Beer, M. Muller, and K. Muhlethaler. (1973) Electron microscopy of selectively stained molecules. In: *Principles and Techniques of Electron Microscopy* 3:55. Hayat, M. A. (ed.). Van Nostrand Reinhold, New York.

Malech, H. L., and J. P. Albert. (1979) Negative staining of protein macromolecules: A simple rapid method. *J. Ultrastr. Res.* 69:191.

Mellema, J. E., E. F. J. Van Bruggen, and M. Gruben. (1967a) An assessment of negative staining in the electron microscopy of low molecular weight proteins. *Biochem. Biophys. Acta* 140:182.

——, E. F. J. Van Bruggen, and M. Gruben. (1967b) Uranyl oxalate as a negative stain for electron microscopy of proteins. *Biochem. Biophys. Acta* 140:18.

Muscatello, U., V. Guarriera-Bobylera, and P. Buffa. (1972a) Configurational changes in isolated rat liver mitochondria as revealed by negative staining. I. Modification caused by osmotic and other factors. *J. Ultrastr. Res.* 40:215.

——, V. Guarriera-Bobylera, and P. Buffa. (1972b) Configurational changes in isolated rat liver mitochondria as revealed by negative staining. II. Modifications caused by changes in respiration state. *J. Ultrastr. Res.* 40:235.

——, V. Guarriera-Bobylera, I. Pasquali-Ronchetti, and A. M. Ballotti-Ricci. (1975) Configurational changes in isolated rat liver mitochondria as revealed by negative staining. III. Modifications caused by uncoupling agents. *J. Ultrastr. Res.* 52:2.

—— and R. W. Horne. (1968) Effect of the tonicity of some negative staining solutions on the elementary structure of membrane-bound systems. *J. Ultrastr. Res.* 25:73.

Pease, D. C. (1964) *Histological Techniques for Electron Microscopy.* Academic Press, New York.

Quintarelli, G., R. Zito, and J. A. Cifonelli. (1971) On phosphotungstic acid staining. *J. Histochem. Cytochem.* 19:641.

Racker, D. K. (1978) Negative staining of whole cells: Transmission electron microscopy of peripheral organelles in rat venous endothelial cells. *J. Histochem. Cytochem.* 26:417.

Silva, M. T., F. C. Guerra, and M. M. Magalhaes. (1968) The fixative action of uranyl acetate in electron microscopy. *Experientia* 24:1074.

Silverman, L., and D. Glick. (1969) The reactivity and staining of tissue proteins with phosphotungstic acid. *J. Cell Biol.* 40:761.

Terzakis, J. A. (1968) Uranyl acetate, a stain and a fixative. *J. Ultrastr Res.* 22:168.

Valentine, R. C. (1961) Contrast enhancement in the electron microscopy of viruses. *Adv. Virus Res.* 8:287.

—— and R. W. Horne. (1962) An assessment of negative staining techniques for revealing ultrastructure. In: *The Interpretation of Ultrastructure* 1:263. Harris, R. J. C. (ed.). Academic Press, New York.

Van Bruggen, E. F. J., E. H. Wienbengu, and M. Gruber. (1960) Negative staining electron microscopy of proteins at pH values below their isoelectric points. Its application to hemocyanin. *Biochem. Biophys. Acta* 42:171.

Watson, D. H. (1962a) Electron micrographic particle counts of phosphotungstate sprayed viruses. *Biochim. Biophys. Acta* 61:321.

—— (1962b) Particle counts of herpes virus in PTA negatively stained preparations. *Proc. 5th Int. Conf. EM* 2:X4.

——, W. C. Russell, and P. Wildy, (1963) Electron microscopy particle counts on herpes virus using the phosphotungstate negative staining technique. *Virology* 19:250.

Watson, M. L. (1958) Staining of tissue sections for electron microscopy with heavy metals. *J. Biophys. Biochem. Cytol.* 4:475.

6. Shadow Casting

A method unlike staining for enhancing contrast is the shadowing or shadow casting technique. Using a vacuum bell jar, a metal is heated to above melting, and the evaporated atoms coat anything in their path (Figure 6-1). If the specimen is held at an angle relative to the evaporation source, the metal will obliquely deposit over the specimen covering only part of it–hence the name "shadow casting." The effect is analogous to shining a flashlight on an object; the object deflects some of the light, which will then reappear some distance behind the object.

Shadowing was originally introduced solely to increase contrast in EM images (Williams and Wycoff, 1946); Bradley (1967) then proved that shadowed particles can also be accurately sized. Another advantage of shadowing is that particles appear in bas-relief, much like scanning electron microscope images. Added advantages of shadowing include support film and specimen stabilization against radiation damage, especially by secondary mechanisms such as heat. The metal film is in contact with the metal grid, and direct contact to ground is thus made.

METHOD OF SHADOWING

The vacuum bell jar is the apparatus used for shadow casting. The basic instrument was discussed in chapter 3 (see "Carbon Support Films") and will not be repeated here; Figure 6-1 shows the configuration for shadowing. The two basic differences operational here are that the carbon electrodes are replaced by a tungsten basket, and

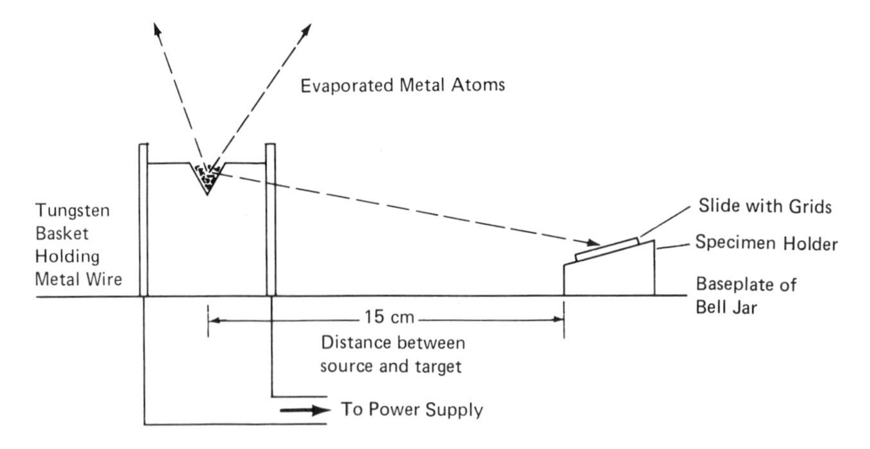

Figure 6-1. Configuration of vacuum bell jar for shadowing.

that the target (sample) is not beneath the source. All other factors (e.g., a diffusion pump and rotary pump in series) are kept constant.

Tungsten is a very high-melting-point metal (3396°C), melting well above the metals used for shadowing. Thus it has proved useful as a substrate to hold the metal for evaporation. The wire is 0.5–1.0 mm in diameter; tungsten baskets are commercially available, but this expense can be avoided if tungsten wire and a wood-screw are available. Simply take about 10 cm of wire and wrap it around the screw threads, having roughly equal amounts of wire on each side of the tip of the basket. Slight heating of the wire in a Bunsen burner makes it more malleable (Bradley, 1967). Excess wire may then be removed

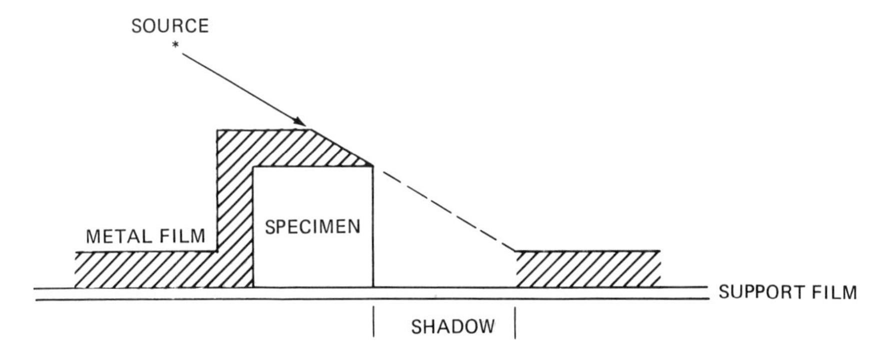

Figure 6-2. A shadowed particle.

and the basket mounted in the high voltage leads. Although the basket becomes brittle with use, if it is unbroken it may be repeatedly used. If the basket breaks, it must be replaced in order to pass a current. The required wire thickness depends upon the current that will be passed through it; higher currents require thicker (1 mm) diameter tungsten wire baskets.

The target (i.e., grids with specimens) is located 10–15 cm from the source at an angle. This separation is required to avoid damage from the heat generated during evaporation, where temperatures as high as 1800°C may be reached. Zingsheim et al. (1970), Rowsowski and Glider (1977), and Slayter (1980) have evaluated this problem.

Oblique deposition of the evaporated film is obtained by mounting the grids on an angled substrate (Figure 6-1). A block of metal or some other material unaffected by high vacuum has its surface cut to a known angle; a number of different blocks and angles (25°, 30°, 35°) are prepared for use ad infinitum. Unless grids are secured, they may be displaced by suction as the rotary pump valve is opened, or by vibrations from the same. Consequently, the grid edges should be secured by double-stick tape to a porcelain plate, which may then be used to indicate film thickness. Fisher (1972) designed a holder for carbon-coating grids; it is also useful for shadowing.

A 10–15-mm length of the metal to be evaporated (e.g., 0.25 mm diameter platinum) is coiled and placed in the tungsten basket. The specimen grids, which have been kept under cover, are placed on the porcelain plate and positioned in the bell jar. As with carbon films, place a drop of oil on the plate for thickness estimation; be sure that the oil does not contaminate the grids. A minimum vacuum of 10^{-4} Torr is required; for reasons to be discussed, the higher the vacuum, the better the thin film. The following information on thin film formation was compiled from Mayer (1955), Bassett et al. (1959), Holland (1970), Maissel (1966), Bradley (1967), Berry et al. (1968), Shiftlett (1968), Neugebauer (1970), Abermann et al. (1972), and Nagatani and Saito (1974).

If a current is passed through the basket, the temperature will rapidly rise in the contained metal wire. When white hot (near the melting point), metal atoms will vaporize and travel away from the source. This is referred to as resistance heating, and applies to non-refractory metals (i.e., those having a melting point less than 2000 K).

The evaporated atoms travel along a straight line because of the high vacuum, until they strike the inside perimeter of the bell jar or its contents. Because a great deal of energy has been transferred to the metal atoms (thermal energy), the atoms will migrate on the interception surface until they reach thermal accomodation with the surface. The atoms will continue to move for a short period of time until one of the following situations is fulfilled:

1. The excess thermal energy is rapidly disseminated by the substrate; for example, the bell jar is considerably cooler than the tungsten basket, and rapidly traps the vaporized atoms (i.e., rapid energy transfers). Thus, cooling the specimen will reduce surface mobility of the vaporized atoms.
2. The metal atoms may be trapped by a binding site on the sample surface; according to Henderson and Griffiths (1972) a high-melting-point metal has strong internal binding characteristics and thus tends to return to its most stable state rapidly.
3. Sufficient vaporized atoms come in contact with one another to form a cluster, referred to as a nucleus in thin film terminology; these nuclei continue to grow as more metal is evaporated, and over time enough clusters are in contact to form a continuous thin film.

Bradley (1958a,b, 1959, 1960, 1967) determined that another way to decrease surface mobility is to simultaneously evaporate carbon (m.p. 3800°C) and platinum (m.p. 1755°C). Apparently the carbon atoms interfere with the formation of platinum crystals; rather, a cluster of carbon–platinum results. Harris (1975) devised a method of electrode arrangement that reliably maintains the ratio of vaporized Pt and C.

A number of attributes of thin films grown by resistance heating become apparent. First, the film must be continuous (i.e., not isolated clusters), but it should not be so thick that it obscures specimen details (Abermann et al., 1972). Consequently, for a given specimen the shadowed film will influence contrast. Ideally, the film thickness should be minimal, but the film must still be continuous. This objective is achieved by reducing granularity (i.e., the size of the isolated clusters). The presence of more but smaller clusters per unit area is strictly a function of the melting point of the evaporative metal and the following sequence is evident: the higher the melting point, the smaller the clusters, the thinner but still continuous the film, and finally the higher the resolution obtainable. Table 6-1 lists the characteristics of the metals (and carbon) typically used in microscopy; refractory metals have been considered but require special equipment (Bachmann, 1962; Hart, 1963; Maissel, 1966; Nagakura et al., 1966; Bradley, 1967; Hintermann and Begnin, 1967; Moor, 1970; Abermann et al., 1972; Abermann and Saltpeter, 1974; Shved and Lylo, 1978).

Gold, with a melting point of 1063°C, is the lowest-melting-point metal used for shadowing. Its density, however, is quite high (Table 6-1). Evaporated films of gold are quite coarse (i.e., the film consists of large grains), and consequently must be quite thick (50–75 Å) to be continuous. At the other end of the spectrum, platinum has a much higher melting point (1755°C), and exhibits minimal granularity with films as thin as 25 Å. Intermediate resolutions are obtainable with palladium, or mixtures of gold/palladium and platinum/palladium. As mentioned earlier, platinum/carbon is used for very fine films. Basically, mixtures are used to offset the negative characteristics of one element by influencing grain size. The choice of a metal for evaporation is basically a function of the desired resolution. For example, if a straightforward examination of *E. coli* is desired, use gold to prepare the sample; very high-resolution shadowing with platinum is unnecessary. In contrast, if an enzyme is being studied, much higher magnification and resolution are required; use platinum. Having at least a vague idea of the relative magnitude of the specimen will help in selection of the appropriate metal, as well as the appropriate shadowing angle (e.g., refer to Smith and Kistler, 1977; Smith and Ivanov, 1980; Scheele and Borisy, 1978; and Vasilier and Koteliansky, 1979, for relative conditions). A great deal of attention is being directed toward determining exact thin film thickness as an outgrowth of SEM and STEM developments (Roli and Flood, 1978; Broers and Spiller, 1980; Flood, 1980); these data are, of course, equally appropriate for TEM applications.

A few other principles of shadowing are important. In addition to the factors discussed above, the degree of vacuum influences granularity: as a rule, the higher the vacuum, the finer the film, regardless of which metal is used (Holland, 1970). The best resolution values

Table 6-1. Metals for Shadowing.

METAL	ATOMIC NUMBER	DENSITY ($kg/m^3 \times 10^{-3}$)	MELTING POINT (°C)	VAPORIZATION TEMP. AT 1 N m^2	ALLOYING WITH W FILAMENT	GRANULATION AND RESOLUTION
Gold	79	19.3	1063	1465	None	Coarse, 50–75 Å
Palladium	46	12.0	1550	1566	None	Coarse, 50–100 Å
Platinum	78	21.5	1755	2090	Considerable	Very fine, 25 Å
Gold/palladium	–	16.1	–	~1566	None	Coarse, 50–100 Å
Platinum/palladium	–	19.4	–	~2090	Slight	Fine, 25–50 Å
Platinum/carbon	–	–			Not applicable	Fine, 25–50 Å
Carbon	6	2.26	3800	2681	Not applicable	Not applicable

listed in Table 6-1 correspond to the vacuum of 10^{-6}–10^{-5} Torr, whereas lower values are lower vacuums. Of course, a point of no return will be encountered with a given vacuum bell jar; but it is very rewarding to be patient until at least 10^{-5} Torr is reached.

Second, because shadowing is extremely sensitive, aseptic techniques must be employed during specimen preparation. Only very pure, dilute suspensions should be used because too much specimen will interfere with the shadows. Consequently, one does not have the freedom of diluting the specimen, for example, by floating it on a drop of stain. It is thus important to prepare several grids with different concentrations to determine the optimum (see Gregory and Pirie, 1973).

Third, a white-hot source is extremely bright. Never look directly at the tungsten filament during heating; protect your eyes with cobalt glass. Alternatively, dependable thicknesses can be determined by performing a trial of current over time; then simply match these values for evaporation.

Following each evaporation, the bell jar should be thoroughly cleaned with acetone and lint-free papers. This is important because immediately after coming to atmosphere, water vapor and gas will deposit throughout the interior. If evaporation is conducted without cleaning, a sandwich of metal–vapor–metal, etc., will build up. A point will quickly be reached at which persistent outgassing of the trapped vapor makes it impossible to reach high vacuum. Along these same lines, when in frequent use the bell jar should be kept under high vacuum.

Finally, the basic advantages of enhancing contrast by shadowing rather than negative staining are that shadowing is aseptic because of high vacuum, and high resolution is readily obtained. On the other hand, the disadvantages of shadowing are that it is time-consuming, and the possibility of radiation damage exists. A primary mechanism of damage is the possible effect of impinging atoms peppering the specimen, and a secondary mechanism is that of heat damage (Rowsowski and Glider, 1977).

REFERENCES

Abermann, R., and M. M. Saltpeter. (1974) Visualization of desoxyribonucleic acid by protein film adsorption and tantalum–tungsten shadowing. *J. Histochem. Cytochem.* 22:845.

——, M. M. Saltpeter, and L. Bachmann. (1972) High resolution shadowing. In: *Principles and Techniques of Electron Microscopy* 2:197. Hayat, M. A. (ed.). Van Nostrand Reinhold, New York.

Bachmann, L. (1962) Shadow-casting using very high melting metals. *Proc. 5th Int. Cong. EM (Philadelphia)* I:FF-3.

Bassett, G. A. (1967) Vacuum evaporated metal films. In: *Techniques for Electron Microscopy*, 2nd ed., p. 411. Kay, D. (ed.). Blackwell Scientific Pub., Oxford.

—— J. W. Manter, and D. W. Pashley. (1959) The nucleation, growth, and structure of thin films. In: *Structures and Properties of Thin Films.* Neugebauer, C. A., J. B. Newkirk, and D. A. Vermilegea (eds.). John Wiley and Sons, New York.

Berry, R. W., P. M. Hall, and M. T. Harris (eds.). (1968) *Thin Film Technology.* Van Nostrand Reinhold, New York.

Bradley, D. E. (1958a) A new approach to the problem of high resolution shadow casting: The simultaneous evaporation of platinum and carbon. *Proc. 4th Int. Cong. EM* 1:428.

—— (1958b) Simultaneous evaporation of platinum and carbon for possible use in high resolution shadow casting for electron microscopy. *Nature* 181:875.

—— (1959) High-resolution shadow-casting techniques for the electron microscope using the simultaneous evaporation of platinum and carbon. *Br. J. Appl. Phys.* 10:198.

—— (1960) Study of background structure in platinum/carbon shadowing deposits. *Br. J. Appl. Phys.* 11:506.

—— (1967) Replica and shadowing techniques. In: *Techniques for Electron Microscopy*, 2nd ed., p. 96. Kay, D. (ed.). Blackwell Scientific Pub., Oxford.

Broers, A. N., and E. Spiller. (1980) A comparison of high resolution scanning electron micrography of metal film coatings with soft X-ray interference measurements of the film roughness. *SEM, Inc.* 1:201.

Fisher, D. G. (1972) A holder for simultaneous fluid processing or carbon coating of electron microscope grids in lots of ten or more. *Stain Technol.* 47:235.

Flood, P. R. (1980) Thin film thickness measurement. *SEM, Inc.* 1:183.

Gregory, D. W., and B. J. S. Pirie. (1973) Wetting agents for biological electron microscopy. II. Shadowing. *J. Micros.* 99:267.

Harris, W. J. (1975) A universal metal and carbon evaporation accessory for electron microscopy techniques and a method for obtaining repeatable evaporations of platinum–carbon. *J. Micros.* 105:265.

Hart, R. G. (1963) A method for shadowing electron microscope specimens with tungsten. *J. Appl. Phys.* 34:434.

Henderson, W. J., and K. Griffiths. (1972) Shadow casting and replication. In: *Principles and Techniques of Electron Microscopy* 2:151. Hayat, M. A. (ed.). Van Nostrand Reinhold, New York.

Hintermann, H. E., and J. Begnin. (1967) Tungsten shadow casting for electron microscopical specimens. *J. Sci. Instrum.* 44:207.

Holland, L. (1970) *Vacuum Deposition of Thin Films.* Chapman & Hall Ltd., London.

Maissel, L. I. (1966) The deposition of thin films by cathodic sputtering. In: *Physics of Thin Films*, vol. 3. Thun, R. (ed.). Academic Press, New York.

Mayer, H. (1955) *Physik Dunner Schichten.* Teil II, Kap. I: Elementasprozesse beim Schichtaufbau, pp. 1-110. Wissenschaftliche Verlagsgesellschaft, Stuttgart.

Moor, H. (1970) High resolution shadow casting by the use of an electron gun. *Proc. 7th Int. Cong. EM* 1:413.

Nagakura, S., M. Kikuchi, K. Aihara, and S. Oketani. (1966) Vacuum-arc evaporation of highly refractory metals and its applications. *Proc. 6th Int. Cong. EM* 1:267.

Nagatani, T., and M. Saito. (1974) Structure analysis of evaporated films by means of TEM and SEM. *IITRI/SEM*, p. 51.

Neugebauer, C. A. (1970) Condensation, nucleation, and growth of thin films. In: *Handbook of Thin Film Technology*, C-8. Maissel, L. I., and R. Glang (eds.). McGraw-Hill, New York.

Roli, J., and P. R. Flood. (1978) A simple method for the determination of thickness and grain size of deposited film as used on non-conductive specimens for SEM. *J. Micros.* 112:359.

Rowsowski, J. R., and W. V. Glider. (1977) Comparative effects of metal coating by sputtering and by vacuum evaporation on delicate features of euglenoid flagellates. *IITRI/SEM* 1:471.

Scheele, K. B., and G. A. Borisy. (1978) Electron microscopy of metal shadowed and negatively stained microtubule protein: Structure of the 30 S oligomer. *J. Biol. Chem.* 253:2846.

Shiflett, C. C. (1968) Evaporated films. In: *Thin Film Technology*, p. 113. Berry, R. W., P. M. Hall, and M. T. Harris (eds.). Van Nostrand, New York.

Shved, A. D., and V. V. Lylo. (1978) Use of tungsten trioxide for shadowing nucleic acid molecules. *Tsitol. Genet.* 12:70.

Slayter, H. S. (1980) High resolution metal coating of biopolymers. *SEM, Inc.* 1:171.

Smith, P. R., and I. E. Ivanov. (1980) Surface reliefs computed from micrographs of isolated heavy metal shadowed particles. *J. Ultrastr. Res.* 71:25.

—— and J. Kistler. (1977) Surface reliefs computed from micrographs of heavy metal shadowed specimens. *J. Ultrastr. Res.* 61:124.

Vasilier, V. D., and V. E. Koteliansky. (1979) Freeze drying and high resolution shadowing in electron microscopy of *Escherichia coli* ribosomes. *Methods Enzym.* 59:612.

Williams, R. C., and R. W. G. Wyckoff (1946) Applications of metallic shadow casting to microscopy. *J. Appl. Phys.* 17:23.

Zingsheim, H. P., R. Abermann, and L. Bachmann (1970) Shadow casting and heat damage. *Proc. 7th Int. Cong. EM* 1:411.

7. Fixation and Embedding

INTRODUCTION

Studying the internal morphology of cells and tissues is preceded by treatment that preserves the specimen as it existed in life as well as protects it against the hostile environment of an electron microscope. Furthermore the specimen must be sliced thin and its density increased for optimal interaction with the electron beam. Thus, the following sequence applies to preparing biological specimens for electron histology (Table 7-1):

1. Fixation: simultaneously sacrificing and preserving the specimen.
2. Dehydration: substitution of cellular water with an organic solvent miscible with water and the embedding medium.
3. Infiltration: saturation of the specimen with the monomer (liquid) form of the embedding medium.
4. Polymerization: curing of the embedding medium into a solid matrix supporting the specimen but which is sufficiently elastic to be cut into thin sections (400–800 Å thick).
5. Ultramicrotomy: sectioning the encapsulated material to a thickness penetrable by the electron beam and mounting it on grids; thick sections ($\sim 2\ \mu m$) may be cut, chromatically stained, and examined by the light microscope.
6. Post-staining: increasing the electron density of the specimen by reaction with heavy metal stains.

Table 7-1. Preparatory steps in electron histology.

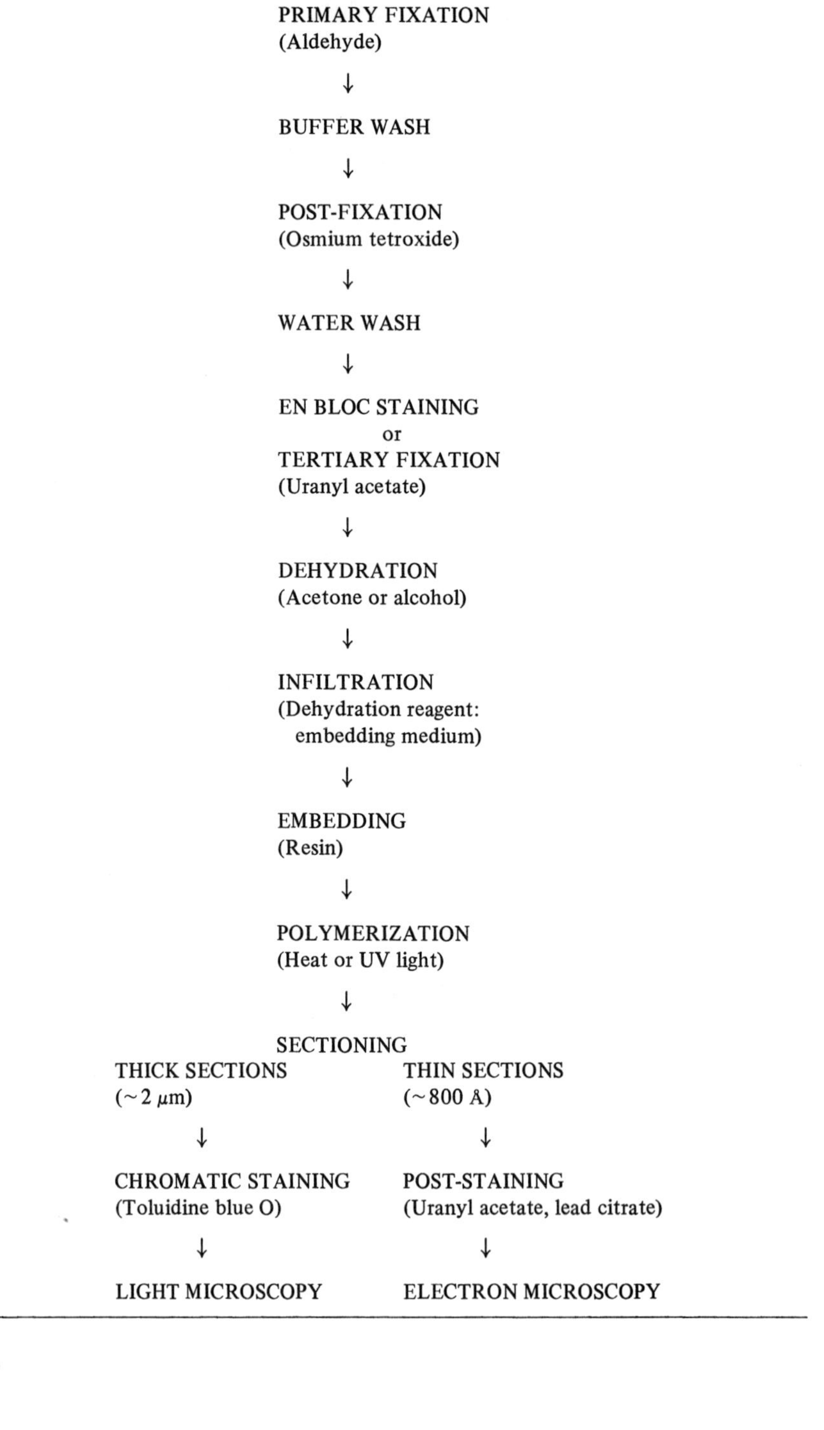

PRIMARY FIXATION
(Aldehyde)

↓

BUFFER WASH

↓

POST-FIXATION
(Osmium tetroxide)

↓

WATER WASH

↓

EN BLOC STAINING
or
TERTIARY FIXATION
(Uranyl acetate)

↓

DEHYDRATION
(Acetone or alcohol)

↓

INFILTRATION
(Dehydration reagent: embedding medium)

↓

EMBEDDING
(Resin)

↓

POLYMERIZATION
(Heat or UV light)

↓

SECTIONING

THICK SECTIONS ($\sim 2\ \mu m$)	THIN SECTIONS (~ 800 Å)
↓	↓
CHROMATIC STAINING (Toluidine blue O)	POST-STAINING (Uranyl acetate, lead citrate)
↓	↓
LIGHT MICROSCOPY	ELECTRON MICROSCOPY

CRITERIA FOR GOOD PRESERVATION

A biological sample should be prepared in such a manner that optical and electron microscopes cannot resolve the artifacts inherent in any given specimen preparation technique, or that the artifacts can be reasonably interpreted in the final analysis. This ambiguous statement is based upon the currently accepted criteria for good preservation. By comparing the effects of physical and chemical electron microscope preparation techniques with data from other fields (e.g., X-ray diffraction), a number of prerequisites have been assembled. These criteria can be separated into three groups: membrane continuity, absence of empty spaces, and comparison with life. A cell may be defined as a system of continuous membranes responsible for ordered energy acquisition, storage, and release. Conversely, a dead cell is observed to have broken membranes. Light and phase-contrast microscopy of living cells demonstrates that membrane continuity is true for at least the more resolvable structures; by logic, it may be assumed that membrane continuity should also exist at the ultracellular level. Likewise, if discontinuity is apparent in fixed tissue at the light microscope level, it may be safely assumed that the ultrastructure also will reflect that discontinuity.

Akin to this criterion is the correlation of electron microscope data with that obtained from instruments other than a microscope. X-ray diffraction is an excellent tool for comparisons of periodic structures (e.g., crystalline inclusions or collagen): if the periodicity is the same in both cases, it may be assumed that fixation is satisfactory.

Another criterion for good fixation is that the cell should not exhibit any empty spaces. In life, the cytoplasm and nucleoplasm are filled with an aqueous solution of protein colloids, carbohydrates, lipids, and so on. Vacuoles are really the only organelles that may appear empty, provided that good preservation of the other organelles exists. In an acceptable transmission electron micrograph, the cytoplasmic ground substance should appear gray, not white (i.e., colorless and therefore void). This parameter is clearly demonstrated when glutaraldehyde/osmium tetroxide–fixed tissue is compared with potassium permanganate–fixed tissue. Fixation with the former results in a dispersed, amorphous, gray precipitate with organelles suspended within it, the precipitate being composed of the reaction products of glutaraldehyde and the protein sol. In comparison, the organelles in permanganate-fixed cells appear to be suspended in a void; the cytoplasmic matrix is unfixed and therefore leached out during subsequent treatments.

CHEMICAL FIXATION

The chemical fixatives used in electron microscopy are organic reagents that promptly react with cellular constituents, simultaneously preserving and stabilizing ultrastructure. Fixatives are thus additive in nature—they become part of the stabilized structure. EM fixatives are also noncoagulant, converting proteins into transparent gels (cf. ethanol, a coagulant fixative that denatures proteins by conversion into reticular solids; see O'Brien et al., 1973).

Proteins constitute approximately 50% of the dry weight of a cell, and much of the protein content serves in a structural fashion. Therefore, it is imperative that proteins be stabilized prior to the onset of autolysis. Because aldehydes are unsurpassed in their ability to cross-link proteins, an aldehyde is employed as the primary fixative. However, other macromolecules may not be stabilized during aldehyde fixation (Pentilla et al., 1974; Mersey and McCully 1978; Demsey et al., 1978); secondary or post-fixation is necessary. The heavy metal stain/fixative osmium tetroxide stabilizes the unsaturated lipids in a cell. As a result of double fixation (i.e., protein and lipid stabilization by reaction with aldehyde and osmium tetroxide, respectively), cell membranes are preserved.

Concurrent with ultrastructure stabilization is separation of the liquid from the solid phases of the protoplasm (Hayat, 1970), thus avoiding translocation of organelles and macromolecules. Essentially, the cell is converted from a gel-like colloid into a spongy network with suspended organelles: the threads of the sponge are precipitated macromolecules. As this real-time event progresses, osmotic pressure differences develop, accompanied by a severe drop in pH. Consequently, fixatives must be buffered and made isotonic relative to the cell interior. If the pH and tonicity are not maintained, the fixative will distort the fine structure.

Other important factors affecting the quality of preservation are the rate of penetration and the rate of fixation, which basically are characteristic of the structure of the fixative molecules. For example, formaldehyde (CH_2O; molecular weight 30) is a much smaller molecule than glutaraldehyde ($C_5H_8O_2$; molecular weight 100.12), and rapidly enters a cell. Glutarladehyde is a dialdehyde with two reactive sites per molecule, while the monoaldehyde formaldehyde has only one reactive site; so in terms of molar ratios, glutaraldehyde will fix much more rapidly than CH_2O. Fixatives that are very strong oxidizers (e.g., acrolein) quickly penetrate and react simultaneously. Along similar lines, the fixative bonds must be unbreakable by subsequent treatment (i.e., the tissue should be inert to subsequent treatment). These characteristics will be expanded upon later in this chapter.

A major parameter influencing those fixative characteristics for a given tissue is the method of fixation (e.g., Johnston et al., 1973; Rosene and Mesulan, 1978). Vascular perfusion, immersion, and in vivo fixation are methods for introducing a fixative to the tissue. The perfusion method of fixation involves substitution of the normal circulatory fluids with the fixative. Because intimate contact between the fixative and cells via the vascular system is established, tissues are rapidly and homogeneously preserved. During in vivo fixation, the tissue is flooded with fixative immediately following anesthesia but while still in the animal. When performed rapidly, in vivo fixation will simultaneously kill and preserve the tissue. When a specimen is removed and then placed in a fixative, immersion fixation occurs; for example, human biopsy specimens will be permeated with the fixative over time. Thus, the method of fixation has a great deal of influence over penetration rates.

These topics and their interrelationships will be discussed in this chapter. More comprehensive reviews and methodologies are presented by Hopwood (1969a), Glauert (1967, 1975), and Hayat (1970). The fixatives used in microscopy are toxic; hence suitable references and handling methods are included in the discussion of each chemical. General lab safety in EM is very well discussed by Humphreys (1977) and Thurston (1978).

BUFFERS

Regardless of the type of fixative used for preservation, a buffer must be included for artificial maintenance of pH and osmolarity of any tissue (a buffer being a solution of a weak acid or base and its salt that resists changes in hydrogen ion concentration when small amounts of a strong acid or base are added to it). Living tissues possess a natural buffering system, but this is overwhelmed by the action of the fixative.

The application of an unbuffered fixative to a cell results in a drastic pH drop. Claude (1961) and Malhotra (1962) noted that after 48 hr of osmium tetroxide fixation, the pH dropped from 6.2 to 4.4. A drop in pH, whether it is due to autolysis or fixation, breaks large macromolecules into smaller subunits. The molecular weight of proteins depends on pH, and proteins impart the characteristic structure of a cell; consequently, radically different, abnormal structures are observed when pH is not controlled (Wrigglesworth and Packer, 1969). For example, autolysis proceeds most rapidly at low pH; if autolytic enzymes escape fixation, they will destroy the unfixed cellular content. Buffers accommodate the hydroxyl and hydrogen ions generated by the reaction between a fixative and macromolecule, and also prevent swelling or shrinkage (osmotic pressure effects): the latter may be controlled by the ionic composition of the buffer or by the addition of non- or weak electrolytes (Boyde and Vesely, 1972; Igbal and Weakley, 1974; Pexieder, 1976).

Although numerous buffers are available, only a few meet the criteria that the ionic composition and pH be within physiological range. In very general terms this range is 6.8–7.1 for plants, and 7.0–7.4 for animals (Butler et al., 1967). The measurement of cytoplasmic pH (~7.0, animals) and organelle pH (e.g., animal nuclei, ~7.7) is extremely difficult, because cells are not static (i.e., pH may vary from moment to moment within a given cell). Consequently, "physiological range pH" for a given tissue is the norm (Glauert, 1975; Schiff and Gennaro, 1979a,b).

Similarly, modifying the tonicity of the buffer/fixative is difficult, because little is known about the natural osmotic pressures of cells (Thornwaite et al., 1978). Even after primary fixation with an aldehyde, membranes are permeable (Feder, 1960; Fahimi and Drochmans, 1965a; Moss, 1966; Favard et al., 1966). Osmotic effects are illustrated in Figure 7-1: isotonicity does not change cell dimensions, hypertonicity causes shrinkage, and hypotonicity swells the cell. Tonicity may be accommodated by the ionic composition of the buffer, by the addition of electrolytes (e.g., NaCl; Rhodin, 1954), or by the addition of nonelectrolytes (e.g., sucrose; Caulfield, 1957; Hampton, 1965; Moretz et al., 1969b: or glucose; McLean,

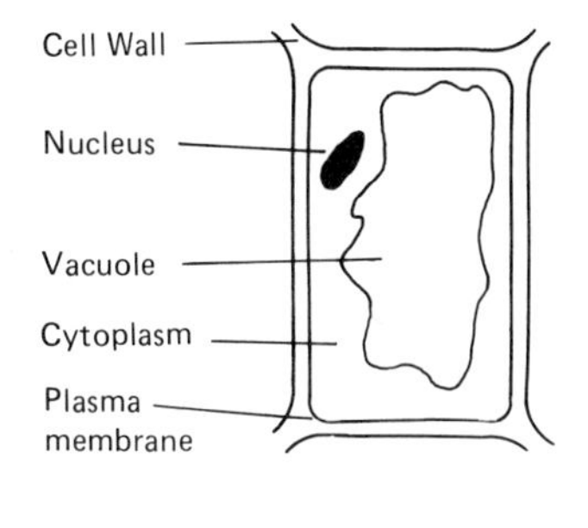

Normally the cytoplasm is pressed as a thin layer along the plasma membrance by the vacuole, and exhibits a relatively high internal osmotic pressure.

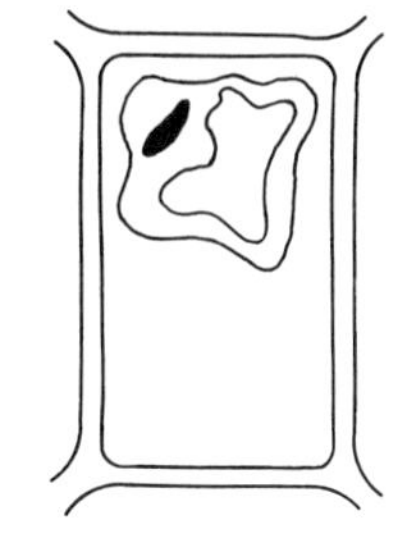

When the cell is placed in a hypertonic solution, water leaves the vacuole and cytoplasm, and the cytoplasm shrinks away from the cell wall (plasmolysis).

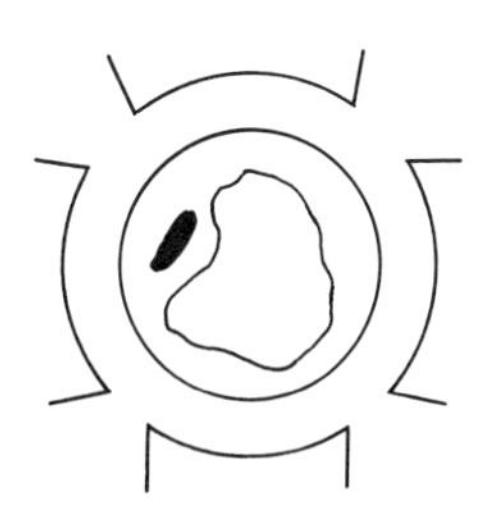

When placed in the hypotonic solution, water may enter the vacuole and cytoplasm, and the entire cell swells.

Figure 7-1. Osmotic effects in a plant cell.

1960). Except for isolated cells, Glauert (1975) suggests that trial and error should be used in evaluating the effects of tonicity: neither swelling nor shrinkage should be visible. She recommends, with some hesitation, the use of an osmometer when isolated cells are studied (Tahmisian, 1964; Maser et al., 1967). The reader may appreciate the extent of this problem by Hayat's (1970) observation that blood plasma is isotonic to 0.31–0.34 M sodium chloride, while kidney cortex is isotonic to 0.23 M NaCl. Furthermore, the fixation effect may be such that cell membrane properties are affected (Carstensen et al., 1971; Bone and Denton, 1971; Hopwood, 1972).

Observations by various authors are, however, slowly adding to our knowledge. Bone and Ryan (1972) as well as Brunk et al. (1975)

systematically evaluated tonicity for given specimen types. Other generalizations concerning electrolytes are that they prevent swelling and lessen extraction, and that divalent cations are 10 to 100 times more effective than monovalent cations (Verweey and Overbeck, 1948; Tooze, 1964). Millonig and Marinozzi (1968) recommend that monovalent cations be used to avoid excess protein precipitation; it is possible that the excess ions present on the divalent species overreact. The addition of electrolytes to a buffer can influence preservation because buffers are active during fixation (Trump and Ericsson, 1965); too little buffering and/or unbalanced ionic composition results in poor preservation. This last problem may be avoided by using nonelectrolytes, which decrease the fixative penetration rate (Hagstrom and Bahr, 1930), and increase extraction (Wood and Luft, 1963 and 1965; Millonig, 1964). Tobin (1980) has concluded that electrolytic buffer ions themselves are most effective. Many types of buffers have been proposed for use in EM (Hayat, 1970; Schiff and Gennaro, 1979a,b), but only a limited number have proved useful for general fixation. Sodium phosphate buffers (Gomori, 1955; Dawson et al., 1969) are closely related to the cell's natural buffer; many different formulations are available and are discussed below under "Glutaraldehyde." A major advantage unique to sodium phosphate is that it can be used for both primary and post-fixation. Sabatini et al. (1963) presented cacodylate for buffering aldehyde fixatives; unfortunately it contains arsenic and requires careful handling. Other buffers include sym-collidine (Gomori, 1946), which may be used as a buffer for osmium tetroxide post-fixation; veronal-acetate buffer (Palade, 1952) was used early in microscopy, but is not acceptable according to current standards (Glauert, 1975).

Kuran and Olzewska (1977) and Schiff et al. (1976) evaluated the degree of macromolecular extraction as a function of buffer type; and Schiff and Gennaro (1979a,b) have comprehensively summarized these data. Good et al. (1966) and Hayat (1970) also provide information on various buffers.

ALDEHYDES

The highest degree of protein stabilization is achieved when aldehydes are used for fixation. They form both intra- and intermolecular crosslinks with the amino acids arginine, cysteine, glutamine/asparagine,

Table 7-2. Aldehydes Common in EM.

ALDEHYDE	CHEMICAL STRUCTURE	MOLECULAR WEIGHT	EFFECTIVENESS FOR EM	SPECIAL CHARACTERISTICS
Formaldehyde	$H_2C{=}O$	30.03	Poor	Rapid penetration
Acrolein	$H_2C{=}CH{-}CHO$	56.06	Good	Potent oxidizer
Glutaraldehyde	$OHC{-}CH_2{-}CH_2{-}CH_2{-}CHO$	100.12	Excellent	Dialdehyde

histidine, lysine, tryptophan, and tyrosine (Bowes and Cater, 1968; Hopwood et al., 1970). The degree of reaction is largely a function of the aldehyde used, as well as the duration of fixation, pH, temperature and concentration. The characteristics of the three most common aldehydes–glutaraldehyde, formaldehyde, and acrolein–are listed in Table 7-2.

Tissue-bound aldehyde groups are referred to as carbonyl compounds, and may be detected by the application of Schiff's reagent (a mild basic fuchsin). In a Schiff-positive reaction, tissues are stained a pale pink as observed by light microscopy. This technique has been modified for electron microscopy using silver as a selective stain for proteins and mucopolysaccharides (Marinozzi, 1961).

Another attribute shared by the aldehydes is that they can rapidly and homogeneously fix large pieces of tissue (2 cm^3); unfortunately uniform post-fixation with osmium tetroxide requires much smaller blocks ($\leqslant 1$ mm^3). Consequently, tissues should be minced into small cubes during aldehyde fixation. Aldehydes are also used in EM cytochemistry (Hayat, 1973), and when various selective staining techniques are to be studied (Hayat, 1975). Mersey and McCully (1978) elegantly illustrate the effects of fixation over time. On the other hand, macromolecules not fixed will be extracted during dehydration and embedding. Furthermore, electron density is not enhanced by low-molecular-weight aldehydes. Consequently, post-fixation with the heavy metal stain/fixative osmium tetroxide is necessary.

Glutaraldehyde

Glutaraldehyde (glutaric acid dialdehyde, $C_5H_8O_2$) is the best protein fixative for electron microscopy (Sabatini et al., 1963; Hopwood, 1972). It is a dialdehyde, meaning that is has two aldehyde groups (R—CH=O) per molecule, one or both of which are available for reaction (Bowes and Cater, 1966). In comparison to the monoaldehydes, although the real molar values or concentrations may be identical, the presence of two aldehyde linkages very effectively increases the overall number of potential reactive sites without an increase in molar concentration.

At the ultrastructural level, Morre and Mollenhauer (1969) showed that membranes are well-stablized with glutaraldehyde, although some membrane reorganization occurs (Jost et al., 1973; Pladellorens and Subirana, 1975; Breathnach and Martin, 1976; Demsey et al., 1978). Excellent preservation of macromolecules has been proved with protein crystal reactions (Quiocho et al., 1967; Moretz et al., 1969b); methods independent of electron microscopy (e.g., X-ray diffraction) have shown that glutaraldehyde induces the least amount of protein conformational changes (Moretz et al., 1969b). This indirectly provides evidence that the size of the glutaraldehyde molecule is such that it perfectly fits between the amino groups of polypeptide chains (Bowes and Cater, 1966). The macromolecules are irreversibly fixed and stable against dehydration and embedding (Quiocho et al., 1967; Richards and Knowles, 1968; Moretz et al., 1969b; Hayat, 1970). Hopwood et al. (1970), Alexa et al. (1971) and Chisalita et al. (1971) investigated the reactivity of various amino acids with glutaraldehyde; refer to this literature or the reviews by Hopwood (1970a,b, 1972 and 1975) for specific reactions.

Glutaraldehyde forms both intra- and intermolecular methylene bridges between reactive side chains of proteins (Habeeb and Hiramoto, 1968) without disrupting peptide bonds (which would result in drastic conformational changes at the macromolecular ⟶ ultrastructural level). Little attention has been directed toward positively identifying the reaction products (Hopwood et al., 1970), but various authors agree that the following sequence is probable:

1. In the presence of a dilute acid or base, two or more molecules of glutaraldehyde "polymerize" by an aldol condensation (Bowes and Cater, 1966; Richards and Knowles, 1968):

$$2(OHC-C_3H_6-CHO) \xrightarrow{OH^-} OHC-C_3H_6-CH{=}C-C_2H_4-CHO$$

Further condensation with aditional glutaraldehyde molecules will yield an aldol having the following structure:

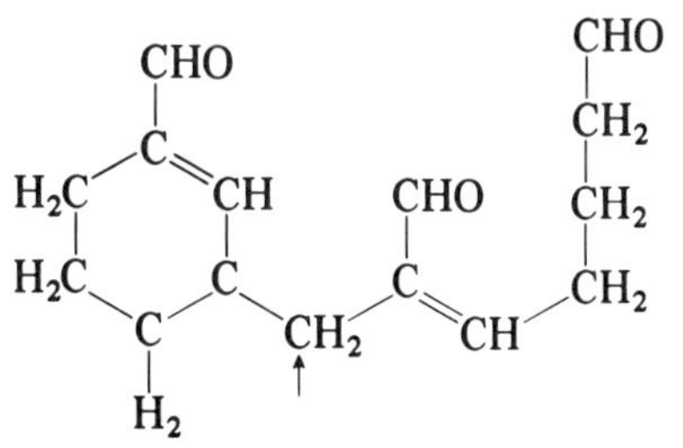

```
              CHO                           CHO
               |                             |
               C                            CH2
             /   \\                           |
        H2C        CH        CHO            CH2
         |          |         |              |
        H2C         C         C             CH2
            \\     /   \\     /   \\\\          /
               C        CH2         CH
               |         ↑
               H2
```

Condensation may continue at the site indicated by the arrow; excess polymerization of stock solutions will destroy the efficiency of glutaraldehyde.

2. The product of the above reaction (aldol) forms cross-links via methylene bridges between the nitrogen atoms of amino groups (Richards and Knowles, 1968):

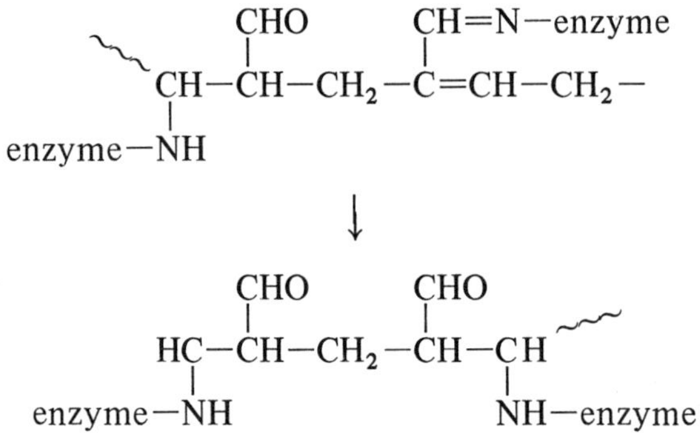

```
                     CHO        CH=N—enzyme
            ~~       |          |
              CH—CH—CH2—C=CH—CH2—
              |
    enzyme—NH

                         ↓

                     CHO        CHO
                     |          |        ~~~
               HC—CH—CH2—CH—CH
               |                 |
      enzyme—NH                  NH—enzyme
```

Jansen et al. (1971) and Tomimatsu et al. (1971) confirmed that stabilization involves two steps: cross-linking preceding polymerization. The precedence of cross-linking over aldol condensation or vice versa most likely is a function of the degree of polymerization within the glutaraldehyde solution.

Habeeb and Hiramoto (1968) and Conger et al. (1978) indicated that other reactive sites are: the N-terminal of some peptides, active hydrogen groups, and amino groups (R—CH=NH). Nonetheless, the cross-linking reaction indicated above results in the most rapid protein stabilization.

Other classes of macromolecules show varying degrees of reactivity toward glutarladehyde. Jones (1972) evaluated the reactivity of glutaraldehyde with unsaturated fatty acids and found selective preservation. Nucleic acids react with glutaraldehyde, but little stabilization results, because only selective amino groups are affected (Hopwood, 1970b, 1972, 1975; Millonig and Marinozzi, 1968). Lipids are predominantly inert (Levy et al., 1965), but are preserved during post-fixation. Hopwood (1967a) showed that 65% of cellular glycogen is retained during glutaraldehyde fixation (vs. 75% retention with formaldehyde); it may be that the glycogen is relatively insoluble, accounting for retention of a portion of it, but some is undoubtedly extracted during dehydration and embedding (e.g., Czarnecki, 1971; Robertson et al., 1975; Ward and Gloster, 1976; Thornell et al., 1977; Sturgess et al., 1978).

The effects of glutaraldehyde at the ultrastructural level must be taken in the context that post-osmication can affect the appearance of organelles. In general, any membranous structure is well-preserved (Morre and Mollenhauer, 1969; Pilstrom and Nordlund, 1975). Mitochondria show good overall preservation, but ultrafine structure may be distorted or absent (Wrigglesworth et al., 1970). Some shrinkage of the cell and nucleus occurs (Hopwood, 1967a; Gusnard and Kirschner, 1977; Skaer and Whytock, 1977; Willison and Rajaraman, 1977), especially during extended exposure to glutaraldehyde. West et al., (1970) noted adequate preservation of chloroplasts. Fawcett (1966) authored an atlas of organelles that thoroughly illustrates organelles and how they were prepared; although the text was probably not intended for this purpose, comparisons of different micrographs strikingly show the effects of different fixatives.

The practical factors that affect the quality of glutaraldehyde fixation are ambient temperature, concentration, and duration of exposure, as well as pH, osmolarity, and type of buffer. Optimal preservation of a given tissue requires that all of these interrelated factors be balanced; in addition to literature searches, often the research microscopist will need to experimentally evaluate several methodologies (e.g., Peracchia and Mittler, 1972; Larsson, 1975; Mersey and McCully, 1978; van Deurs and Luft, 1979).

Fixation is normally conducted at 0–4°C to: retard autolytic activities, ensure uniform fixation, and prevent extraction resulting from overfixation (Langenberg, 1979). Warm temperatures promote fixative as well as autolytic activity, meaning that abnormal structures would be preserved (Chambers et al., 1968). Thus, fixation is always done in the cold; subsequent preparation steps are also at 0–4°C with gradual warming to room temperature in 100% dehydration reagent.

Hopwood (1967a) measured the depth of penetration of 4% glutaraldehyde into liver (24-hr exposure) and reported 4.5 mm and 2.5 mm at, respectively, room temperature and 0–4°C. He concluded that for optimal penetration/fixation depths, the tissue should be limited in size to $\leqslant 1\ mm^3$ for fixation time of about 2 hr. Flitney (1966) recommended that tissue be minced into cubes of $0.5\ mm^3$; the author agrees that smaller blocks should be used because the EM novice may find large specimens difficult to section, and, more important, post-fixation is most uniform with small cubes. Tissues are easily minced with two razor blades, and the tissue is positioned in a plastic Petri dish filled with glutaraldehyde. The razors must be sharp to avoid mechanical deformation; likewise, do not tear the tissue. Cutting in a glass Petri dish is simplified by either cutting on a sheet of parafilm, or preparing a dental wax layer a few millimeters thick.

The fixative concentration and duration of fixation are closely related. Plant or animal tissue cubes maintained at 0–4°C are fixed in 2.0–4% glutaraldehyde within 2 hr (Fahimi and Drochmans, 1965a, 1965b; Schultz and Karlsson, 1965; Pentilla et al., 1974, 1975; Mersey and McCully, 1978). Lipid retention is best at $\sim$3% as shown by various authors (Busson-Mabillot, 1971; Mathieu et al., 1978). Fahimi and Drochmans (1965a) concluded that very high concentration ($\sim$37%) and osmolarity ($>$570 mOsM) will shrink the tissue, whereas low concentration ($\sim$0.5%) and osmolarity ($<$197 mOsM) induce swelling. They, in addition to Hampton (1965), Maunsbach (1966), Arborgh et al. (1976), Deutsch and Hillman (1977), and Thornwaite et al. (1978), have shown that a slightly hypertonic, moderately concentrated solution is optimal. This holds true for most plant and animal tissues; exceptions are cell suspensions such as bacteria or tissue cultures (concentration 0.5–1.5%), which will be discussed later (Buckley, 1973a,b). Chambers et al. (1968) recommend measuring the osmolarity of glutaraldehyde solutions with an osmometer.

The tonicity and pH of the fixative are of course closely related to the buffer vehicle (Fahimi and Drochmans, 1965b, Arborgh et al., 1976; Pexieder, 1976; Mathieu et al., 1978). When buffered, glutaraldehyde alone has little influence on specimen volume changes (Bone and Denton, 1971; Bone and Ryan, 1972; Brunk et al., 1975); in effect, then, slightly hypertonic solutions are actually used for fixation (Bone and Ryan, 1972). The problem of measuring tonicity is again apparent. Phosphate or cacodylate buffers are commonly used with glutaraldehyde, the main criterion for their selection being comparable buffer osmolarities (Busson-Mabillot, 1971). Phosphate buffers are similar to natural physiological fluids, and contain both mono- and dibasic sodium phosphates. Sorenson's phosphate buffer

(Gomori, 1955) shows variable osmolarity from pH 5.8 to 8.0; its osmolarity values at pH 7.2 are as follows (Glauert, 1975):

0.05 M Sorenson's buffer →	118 mOsM
0.075	180
0.10	226
0.15	350

Osmolarity may also be adjusted with sucrose; 0.1 M buffer combined with 0.18 M sucrose results in 425 mOsM (Glauert, 1975). Preparations of this and other buffers are included in Chapter 11.

Various other phosphate buffers have been researched, basically to permit ready adjustment of pH and tonicity. Glauert (1975) comprehensively reviewed phosphates; the following summary is from her text. A slightly hypertonic solution (440 mOsM) at pH 7.4 was used by Millonig (1964) and Karlsson and Schultz (1965) for hydrated tissues. A hypotonic solution (298 mOsM) at pH 7.35 was employed for perfusion of kidney (Maunsbach, 1966), whereas a buffer of pH 7.4 and osmolarity 320 mOsM was used by Karlsson and Schultz (1965) for perfusion of the central nervous system.

Over time, phosphate buffers will support microorganism growth; this effect can be delayed by refrigerating the stock solutions, but it is recommended that fresh solutions be prepared monthly. Some precipitation within prepared cells may also be noticed, as Gil and Weibel (1968) observed with lung. Another advantage of phosphate buffers is that they may be used for post-fixation with osmium tetroxide.

Cacodylate buffers, which yield much the same results as phosphate buffers, were first introduced by Sabatini et al. (1963). Their osmolarity may be adjusted with sucrose, glucose, or NaCl; Brunk et al. (1975) have compared the effects of different osmolarities. Cacodylate buffers must be carefully handled because they contain arsenic; therefore always work in a fume hood (there is also a bad odor) and wear rubber gloves when handling this buffer (Weakly, 1977). Because of its toxicity, the effect of cacodylate on the specimen's cells should not be ignored (Schiff and Gennaro, 1979b). Check the pH of stock solutions before use; even though they are relatively stable and are not contaminated, a pH decrease occurs over time (Glauert, 1975). Another disadvantage of cacodylate, unlike phosphate, is that it cannot be used for buffering osmium tetroxide; therefore a switch to sym-collidine is necessary. Cacodylate will also react with uranyl acetate during en bloc staining (Silva, 1973).

In summary, glutaraldehyde fixation should be used under the following conditions for most tissues:

Temperature 0–4°C
Time 1–2 hr
Concentration 2–4%
Buffer: cacodylate or phosphate
pH: 7.0–7.4
Osmolarity of fixative/buffers solution: ~200–300 mOsM (must be adjusted for tissue type)

Glutaraldehyde is commercially available in aqueous solutions of 25–70% concentration in large volume; smaller 10-ml ampoules are available in 8, 25, 50 and 70% concentrations. Hayat (1970) and Glauert (1975) discuss the purification of biological-grade glutaraldehyde for EM; today, EM-grade fixative that is purified and stored in an inert atmosphere (nitrogen) is available (cf. Jones, 1974; Weakly, 1974; Gillett et al., 1975). Robertson and Schultz (1970) evaluated the effects of impure glutaraldehyde on fixation. When many tissues are to be processed over time, larger volumes of aldehyde should be purchased; if tissues are only rarely fixed, vials of glutaraldehyde are more economical. These stock solutions, as well as the dilute fixative/buffer should be refrigerated and kept in the dark (e.g., amber bottles or foil wrapping) to prevent polymerization. These solutions can be checked for the degree of polymerization by simply combining a few drops of the EM-grade glutaraldehyde with a few drops of water; if a precipitate forms, the solution should be discarded. Likewise, a drop in pH to $\leqslant 4$ indicates polymerization (Aso and Aito, 1962).

Direct contact with glutaraldehyde should be avoided. Always wear gloves and work with open solutions in a fume hood. An allergy to aldehydes may develop with chronic exposure. During the fixation period, vials of sample in fixative may be stored in a refrigerator, provided that the vials are tightly sealed and labeled.

After completion of primary fixation, the tissue is rinsed with several cold washes of the same buffer as used in fixation. This will remove excess unreacted glutaraldehyde from the tissue, which might otherwise react with subsequent reagents. For example, glutaraldehyde will react with osmium tetroxide, resulting in a dense precipitate at the surface and within the tissue block (Trump and Ericsson, 1965; Trump and Bulger, 1966). Washing also reduces the amount of osmium molecules available for reaction with the tissues, and embrittles the tissue to a point where thin sectioning is very difficult. However, Ockleford (1975) referred to washing as redundant, and rapidly prepared good specimens without this intermediate step.

On the other hand, prolonging the buffer wash will result in shrinkage and extraction; cacodylate buffers severely affect some organelles when exposed for as little as 30 min. (Bodian, 1970). In comparison, Glauert (1975) has noted that storage of glutaraldehyde-fixed specimens overnight or even longer is tolerable, provided that the osmolarity of the buffer is altered to fit that of the fixed tissue (native tissue osmolarity may be different from that of fixed tissue; Jost et al., 1973). However, the author recommends that lengthy times between primary and post-fixation should be avoided whenever possible. Dense tissues (e.g., leaves) should be washed for several hours, whereas less dense tissues may be washed from 15–60 min. (e.g., liver vs. skin). Some of these problems may be avoided by using a mixture of glutaraldehyde and osmium tetroxide for primary fixation (Trump and Bulger, 1966); this is a delicate situation because as noted above these two fixatives react with one another. More will be said about combined fixatives in the discussion of osmium tetroxide.

Formaldehyde

The most common, general-purpose fixative used in biology is formaldehyde, CH_2O, with one reactive group per molecule (it is a monoaldehyde) and a very low molecular weight (30.03). Aqueous solutions of formaldehyde (normally a gas) are referred to as formalin; the solid polymer is paraformaldehyde, $(CH_2O)_n$. However, formaldehyde cannot fulfill the criteria for good preservation of ultrastructure, and is used only rarely for electron microscopy.

The reactions between formaldehyde and proteins have been thoroughly investigated in the tanning industry, where it is used for curing of leather (Bowes and Cater, 1966, 1968); evaluations at the fine-structural level have been discussed by Sabatini et al. (1962, 1963), Baker (1965), Baker and McCrae (1966), Hopwood (1969), and Artvinli (1975). The basic cross-linking reaction is a two-step process: first, methylol groups result when CH_2O reacts with free amino groups. Second, a condensation reaction between the methylol groups and phenol, imidazole, and indole results in cross-linking via methylene bridges (Lojda, 1965). Because formaldehyde is not routinely used in EM preparations, its chemistry will not be dwelled upon; the references cited should be consulted for detailed discussions (see also Ealker, 1964).

The reaction between CH_2O and lipids is poorly understood (Wolman, 1955; Wolman and Greco, 1952; Jones and Gresham, 1966); while some modification occurs, most lipids are extracted during dehydration and embedding. Similarly, nucleic acids are modified, but the reaction is reversible (Eyring and Ofenfand, 1967). The major reason why formaldehyde should be avoided in EM is that protein reactions are also readily reversible by hydrolysis (Haselkorn and Doty, 1961; Eyring and Ofengand, 1967).

A few situations may arise, however, in which formaldehyde may be used for primary fixation. Gonzalez-Aguilar (1969) successfully perfused brain, and Winborn and Seelig (1970) noted acceptable preservation when large blocks of tissue ($\sim$3.5 cm^3) are to be fixed. In hospital laboratories, a biopsy initially studied by light microscopy may require electron microscopy. Formalin is typically used in this setting; the microscopist should not balk when faced with preparing formalin-fixed tissues, usually because that is the only available specimen. Subjecting a patient to another biopsy simply to fulfill the criteria for preservation is absurd; therefore prepare the tissue as received, and tolerate the artifacts. The combination of formalin with other aldehydes also has applications which will be considered below under "Comparison of Aldehydes."

Acrolein

Another monoaldehyde useful in EM is acrolein, or acrylic aldehyde (C_3H_4O). Introduced by Luft (1959), it is by far the most powerful oxidizer when compared to other fixatives. Hence, comparatively low concentrations and short exposures of acrolein are necessary for good fixation (Sabatini et al., 1964; Davey, 1973; Saito and Keino,

1976; Landis et al., 1980). Appropriate specimen types have compressed cell walls (e.g., coleoptiles) which are readily penetrated by the small acrolein molecules. Unless precautions are taken when handling acrolein, the user will also be fixed!

Acrolein's very rapid oxidative capacity fixes proteins by reaction with sulfhydryl, aliphatic NH_2 and NH, and imidazole groups (van Duijn, 1961). These sites have been observed with the Schiff reaction (Marinozzi, 1963). Acrolein is soluble in lipids (Norton et al., 1962; Schultz and Case, 1968).

The problem of determining the proper fixation time when using acrolein is a major reason why it is not used for routine tissue preparation. Severe extraction of unfixed macromolecules is initiated immediately after (and possibly during) tissue exposure to the fixative (Sabatini et al., 1963). Consequently, specimens are fixed in very low concentrations (0.5–5.0%) of cacodylate-buffered acrolein for short periods of time (~15 min.). As will be seen, fixative mixtures of acrolein and glutaraldehyde are applicable to a larger variety of specimen types.

Acrolein is a much more hazardous chemical than the other aldehydes; acute effects of self-exposure to acrolein are skin, respiratory tract, and eye sensitization. It is also flammable. Therefore, safety precautions must include working in a fume hood and wearing gloves and goggles. It is also inadvisable to keep acrolein on hand; purchase only as much as needed when it is needed.

Both purified acrolein and stock solutions will polymerize upon exposure to air; refrigerate the fixative in an amber bottle, and purchase purified acrolein in ampoules that have an inert atmosphere. Organic-contaminant-free glassware must be used to avoid reaction between this aldehyde and such contaminants (Albin, 1962); cleaning of glassware with 10% hydrofluoric acid followed by several rinses with distilled water is recommended. Hayat (1970) indicates that solutions are contaminated if turbid or if the pH of 10% acrolein in tap water falls below 6.4. Disposal of acrolein is by combination with 70% sodium bisulfite.

Comparison of Aldehydes

The various aldehydes differ in their rates of penetration, fixation, cross-linking, and stability, as summarized in Table 7-3. Before any fixative can react with macromolecules, it must penetrate the tissue.

Table 7-3. Comparison of aldehydes.

	RELATIVE RATES			
ALDEHYDE	PENETRATION	FIXATION	CROSS-LINKING	STABILITY
Glutaraldehyde	Slow	Moderate	Excellent	Excellent
Formaldehyde	Moderate	Good	Poor	Poor
Acrolein	Very rapid	Very rapid	Moderate	Moderate
Glutaraldehyde/ formaldehyde	Moderate	Good	Excellent	Excellent
Glutaraldehyde/ acrolein	Rapid	Rapid	Excellent	Excellent

Perfusion of tissues containing a great deal of vasculature (e.g., kidney) enhances fixation because the tissue is rapidly infiltrated with the fixative (Chambers et al., 1968); during in situ or in vitro fixation the fixative must diffuse through successive layers of cells. As a result, the method of fixation in practical terms greatly influences effective penetration rates. Slower rates are inherent when the in vitro or in situ methods are employed, and the influence of the fixation rate becomes important. Feder and O'Brien (1968) compared aldehyde penetration rates and determined that acrolein is most rapid, followed by formaldehyde, and glutaraldehyde is slowest. The high oxidizing capacity of acrolein accounts for its rapid penetration, whereas the tiny formaldehyde molecules readily enter cells (e.g., Zeikus and Aldrich, 1975).

Fixation rates follow the same sequence, but the stability of the bonds is very different; glutaraldehyde bonds far exceed the stability induced by acrolein or formaldehyde. Bowes and his research team used collagen to measure the number and stability of cross-links in collagen (Bowes and Kenten, 1949; Bowes and Raistrick, 1964; Bowes and Cater, 1966; Bowes et al., 1965; Trnavska et al., 1966; Alexa et al., 1971).

Comparison of these rates for pure aldehydes indicates that glutaraldehyde is most effective for general fixation, especially in conjunction with post-osmication (Sabatini et al., 1963). In situations where its slow penetration rate is a disadvantage (e.g., very large tissue specimens or dense samples such as pollen), glutaraldehyde mixed with another aldehyde may be used. Karnovsky (1965) found that a mixture of formaldehyde and glutaraldehyde resulted in good preservation; the former rapidly penetrates and reacts, permitting influx of glutaraldehyde, which subsequently displaces the tissue-bound formaldehyde by providing more stable bonds. Similar results using acrolein–glutaraldehyde mixtures have been reported by Sandborn (1966), but the precautions when handling acrolein must be enforced.

METHODS OF PRIMARY FIXATION

Regardless of the type of specimen being studied, fixation should be initiated as soon as possible to avoid autolytic artifacts (Schmalbruch, 1980). Mechanical damage, whether by overhandling or tearing during mincing, should be avoided. The tissue cubes may then be transferred to a small test tube or vial and remain there throughout fixation and infiltration (solutions are decanted). Small receptacles are recommended to avoid excess waste of chemicals; some EM reagents are quite expensive (e.g., the current market value of OsO_4 is $13.00/g).

The three basic methods of fixation are in vivo fixation, immersion (in situ) fixation, and vascular perfusion. In vivo fixation involves introducing the fixative to the living tissue by injection or pouring the solution over the undissected organ of interest; after excision, the tissue is subjected to immersion fixation (i.e., the tissue is submerged within the fixative, minced, then transferred to a vial of fresh fixative). Immersion fixation may also be conducted alone, as that tissue is removed and placed in the fixative. Vascular perfusion introduces the fixative to the organism via its vasculature, thus replacing normal circulatory fluids.

Although there is overlap among methods of fixation for specimens as diverse as plants and bacterial cultures, they are considered separately to avoid confusion. Glauert (1975) extensively lists the various specimen types and treatments. Please refer to Chapter 11 for specific methodologies.

Animal Tissues

Animal tissues may be prepared by any method of fixation; the choice is defined by the type of animal (experimental or human) or the tissue being studied. General anesthetics such as ether or sodium pentobarbital are useful for most applications (Altmann and Dittmer, 1973; Glauert, 1975). Barnes and Etherington (1973) discuss various drug dosages for appropriate experimental animals. Alternatively, sacrificing the animal by spinal dislocation or beheading and immediately opening the desired body cavity also works well.

Bathing the organ of interest with cold fixative before removal from the animal is effective for mechanically stabilizing the tissue (application for 5–10 min.), provided that the area of interest is close to the organ surface (Maunsbach et al., 1962). The tissue is then removed, minced, and fixed for 1.5–2 hr at 0–4°C. Abrunhosa (1972) recommends injecting cold fixative into an organ for deeper penetration.

Immersion fixation alone is used when sacrifice of the animal is not desired. A suitable piece of tissue is removed from the body and placed in chilled fixative. After approximately 15 min. of fixation, the tissue is minced into cubes 0.5 mm on a side, and the fixation is continued for a total of 1–2 hr. If excess blood is observed at the tissue surface, rinse it free with fixative. This avoids potential embrittlement, which makes tissue difficult to thin section.

Although the above technique provides adequate preservation, two situations arise in which vascular perfusion is preferred: (1) oxygen deprivation damages the central nervous system, and the tissue must be infused before damage occurs (Karlsson and Schultz, 1965; Kalimo, 1976); (2) the internal areas of tissue (e.g., lung or kidney) are not readily penetrable by fixatives, and dissection may mechanically damage the tissue (Gertz et al., 1975; Glauert, 1975). Consequently, in these examples vascular perfusion is the best fixation method. This technique is nontrivial, a large degree of its success depending upon the microscopist's skill.

Palay et al. (1962) described a gravity-fed perfusion apparatus that is outfitted with two reservoirs, one for Ringer's solution and the other for fixative. The Ringer's solution (Glauert, 1975, recommends saline) serves as a perfusate to precede the fixative and dilute the blood. Gil and Weibel (1969–70) describe a perfusion pump for preparation of lung and other tissues; also see Baker and Rosenkrantz (1976) for volumetric perfusion of lung. However, for noncritical applications simpler perfusion methods employing only one gravity-fed reservoir of fixative are sufficient. With a gravity-fed system, care must be taken to ensure that the fixative is gently but thoroughly perfused through the organ of interest (Van Harreveld and Khattab, 1968). Glauert (1975) indicates that the reservoir height should be 120–150 cm or 20–30 cm for intra-arterial or intravenous routes, respectively.

After preparing the perfusion apparatus, the animal is anesthetized, its heart exposed, the cannula introduced ~5 mm into the aorta and held in place (in small animals it may be easier to introduce the cannula into the left ventricle), perfusion initiated, and the right atrium cut to allow escape of body fluids (Schultz and Case, 1970). For tissues of the head, clamp the abdominal aorta shut, and sever the jugular and then the inferior vena cava.

The temperature of the fixative is variable; preliminary studies may be conducted at room temperature, but body temperature is more often recommended (Hayat, 1970). Vasoconstriction occurs when the perfusate is lower than body temperature or is hypotonic. A solution of 2–3% buffered glutaraldehyde, osmotically adjusted for the desired tissue (slight hypertonicity is useful), is perfused for about 15 min., the tissue dissected and minced, then immersion-fixed in the same glutaraldehyde solution for 1–2 hr. Post-fixation, and so forth, then follow normal schedules.

Some generalizations concerning the target tissue's natural vascular distribution may also affect ambient conditions for fixation. Maunsbach (1966) obtained good preservation of rat kidney with 0.25% fixative, whereas Schultz and Karlsson (1965) were satisfied when central nervous system tissue was perfused with 2.5% glutaraldehyde.

Plant Tissues

Botanical specimens have an advantage over animal tissues in that they are generally easier to handle physically, except for their high internal osmotic pressure; and, when a dense, waxy cuticle is present, it may resist penetration and require modified treatment. Consequently, although identical reagents are used throughout preparation of plant and animal tissues, considerably longer exposures at each stage are commonly necessary for plant tissues. Lower plants such as algae are prepared by a compromise between the techniques discussed here and those for isolated cells.

Immersion fixation of small cubes or slices of tissue is most commonly conducted. These tissues must be cut without distortion using a sharp razor; the author recommends that pieces be fairly thin for most rapid penetration of reagents. When transferred to a vial of fixative, the tissue has a tendency to float; therefore apply mild negative pressure using an aspirator to assist in fixative penetration by removal of air.

Some tissues (e.g., leaves) may be perfused without damage to the entire plant. The fixative may be injected into a major vascular channel, and the specimen excised, and then further processed by immersion fixation. Alternatively, if the surface of the leaf is of interest (e.g., stomata), a section of the leaf may be isolated by a ring of petroleum jelly. The ring is filled with fixative and excised 10–15 min. later for continued immersion fixation.

Falk (1980) summarizes fixation conditions for plants. Glutaraldehyde and/or acrolein is used for primary fixation. Acrolein alone gave better preservation than a mixture for tissues surrounded by dense walls (Mersey and McCully, 1978); nonetheless, glutaraldehyde (concentration 2–4%) appears more common. The pH of the buffered fixative solution is 6.8–7.2 at tonicity ~0.1 M. Cold fixation is complete within ~2.5 hr, although it may be necessary to increase this time.

The buffer wash between primary and post-fixation is significantly longer for plants than animals, requiring up to 3 hr with numerous buffer changes. Again, osmotic pressure differences between the plant cell interior and buffer must be controlled (Salema and Brandao, 1973). Frequent agitation and changing of the buffer will accelerate the wash. The rotary shakers discussed below under "Embedding" are useful here. Osmium post-fixation is complete within 1.5–2 hr; dehydration times should be increased, and embedding done in a low-viscosity resin (e.g., Spurr).

Isolated Cells and Cell Fractions

Bacteria growing in liquid media, white blood cells, or ribosomes are examples of this classification. There are two closely related methods for preparing cells and cell fractions: (1) the specimen may be centrifuged into a pellet and processed similarly to tissue blocks;

(2) a suspension is concentrated by mild centrifugation, the excess supernatant decanted, and the specimen resuspended in fixative.

Suspended samples that can tolerate rather forceful centrifugation into a compact pellet may be fixed in the centrifuge tube. The tissue is processed exactly like animal tissues using immersion fixation; large pellets should be exposed to the fixative for a few minutes, removed, and minced into small cubes. This technique may be modified by mild centrifugation of the specimen, removal of some of the excess supernatant (but not all—the cells are still suspended), and addition of an equal amount of doubly concentrated fixative: this will dilute the fixative from, for example, 4% to the working concentration of 2% (Glauert and Thornley, 1966). The individual cells thus have immediate reaction with the fixative, an advantage when dense-walled specimens are being studied (e.g., bacterial spores). While in the fixative, the cells are centrifuged, and the pellet is treated as above. Deter (1973) presents an excellent discussion of centrifugation.

A major problem with the above methods is that noncohesive pellets disintegrate during processing, and resuspension/centrifugation may be necessary at each preparation step. Ryter and Kellenberger (1958a) developed an ingenious method for maintaining the integrity of small pellets. They embed pellets in agar, which is permeable but nonreactive toward the fixation and embedding reagents, and the agar blocks are treated as animal tissues. The method is as follows:

1. Prepare 2% aqueous agar and maintain it in the liquid state by warming in a water bath at 45°C. The fixed cells are also warmed to 45°C.
2. Transfer a small drop of agar (use a warm pipette) to the sample test tube, and gently shake it to form a suspension. Compact pellets should be resuspended in a small volume of fixative prior to adding the agar; the addition of too much agar will severely dilute the sample.
3. Immediately transfer the specimen/agar onto a cool microscope slide using a warm pipette or by simply pouring out the contents of the test tube.
4. The agar will solidify within a few minutes; then cut cubes (~0.5 mm^3) of the specimen and continue processing.

Because it is difficult to see the pellet during embedding, a method for staining the agar with basic fuchsin was recommended by Winters and Slade (1971).

Other methods have been developed to maintain pellet integrity but avoid the high temperatures that may adversely affect temperature-sensitive cells and/or, if heating is prolonged, lead to polymerization of the glutaraldehyde molecules. The fibrin clot method (Charret and Fauré-Fremiet, 1967) has the additional advantage that the support matrix may be adjusted from a spongelike mesh to one more dense. Furtado (1970) and Glauert (1975) describe this method.

Another embedding method developed by Shands (1968), uses 2% bovine serum albumin, which "polymerizes" into a gel following exposure to glutaraldehyde. Fixed cells are suspended in the albumin, a few drops of concentrated glutaraldehyde are added, and the suspension is immediately centrifuged. The gel is minced and processed conventionally. Sawicki and Lipitz (1971) used egg-white albumin for encapsulating cells; the polymerization of drops of the albumin/cell suspension is by exposure to concentrated (40%) formaldehyde fumes.

Cultured Cells Growing on a Substrate

The preparation of bacteria growing on agar, organ cultures, and monolayers of cultured cells are essentially identical. Typically, the cells are rinsed with a warm balanced salt solution to remove extraneous matter, and the primary fixative is introduced directly into the vessel. For example, bacterial colonies growing on agar are flooded with fresh fixative (cold or at the normal cell temperature and then cooled), and cubes of the colony/agar are removed with razor blades, transferred to fresh fixative, and processed according to standard methods.

Cultured monolayers of cells are extremely sensitive to changes in osmotic pressure, and it is usually necessary to prepare several specimens using different buffer tonicities. Current literature (Collins et al., 1980; Shay and Walker, 1980) indicates that 0.1 M sodium cacodylate or phosphate buffers are the best starting points. Typical fixation schedules are quicker than for other specimen types, although 2% concentrations still are standard; usually, exposure for 30–60 min. per fixative is accepted (Shay and Walker, 1980). Other special methodologies for tissue cultures are discussed by Nelson and Flaxman (1972), Buckley (1973a), Gorycki and Askanas (1977), Perre and Foncin (1977), Beesley (1978), and Johnson (1978); Nopanitaya et al. (1977) grew cultured cells on polycarbonate filters and processed the filter according to conventional methods.

OSMIUM TETROXIDE

O O
Os
O O

Osmium tetroxide is very effective for stabilizing a variety of macromolecules, most notably unsaturated lipids, and its high molecular weight (254.20) significantly increases a specimen's electron density (Millonig and Marinozzi, 1968). Although early EM researchers employed osmium as the sole tissue fixative, today it is used for secondary (post-) fixation with an aldehyde being the primary fixative; this sequence is referred to as double fixation. The resulting quality of preservation is excellent, far surpassing that of only one fixative (Sabatini et al., 1963; Machado, 1967).

When tissue cubes are exposed to buffered solutions of osmium tetroxide, they promptly begin reacting with the specimen at its surface, resulting in a general increase in tissue density. Over time, as more metal is deposited in the specimen, a barrier will form and prevent continued influx of the fixative. Consequently, the rate of penetration decreases significantly over time (Burkl and Schiechl, 1968), and tissue cubes must be $\leqslant 0.5$ mm^3 in size for thorough infiltration. Because ultrastructure has been partially stabilized during primary fixation, the slower penetration by OsO_4 is tolerable. In addition, the conversion of the cytoplasm from a colloidal into a more solid network (much like a sponge) during aldehyde fixation allows more ready penetration.

Other parameters affecting the rate of OsO_4 penetration are temperature, buffer type and tonicity, and concentration of osmium. Because the fixative is applied by the immersion method, some of the factors discussed for aldehydes (i.e., vascular perfusion, in vivo fixation) do not apply here. Regardless of the vehicle, osmium penetrates more rapidly at room temperature than at 0–4°C (Caulfield, 1957). However, at room temperature fixation is uneven and extraction of unfixed macromolecules is increased (Sjostrand, 1956; Stein and Stein, 1971). Therefore, primary and post-fixation are conducted at 0–4°C.

The vehicles commonly used with OsO_4 are sodium phosphate or collidine buffers. It is known that the ionic composition of a buffer, in addition to reagents used for adjusting tonicity, affect the rate and quality of preservation (Trump and Ericsson, 1965; Hayat, 1970), but many of these studies were conducted using osmium as the primary fixative. Relatively few quantitative data have been collected for double fixation (Glauert, 1975; Schiff and Gennaro, 1979a,b) because it is incredibly difficult to systematically assess the multiple effects of fixation. Nonetheless, some generalizations may be made about buffer vehicles.

Sodium phosphate buffers are amenable to both primary and post-fixation, as discussed earlier (Millonig, 1961). Glauert (1975) recommends using 0.1 M buffer at pH 7.0–7.4 with the addition of sucrose for tonicity adjustment. On the other hand, Hayat (1970) generally uses 0.3 M buffer, but cautions that sucrose additives decrease the rate of penetration (Hagstrom and Bahr, 1960). In general, osmotically balanced solutions penetrate slowly but decrease swelling. Bone and Ryan (1972) conclude that the osmolarity of OsO_4 for post-fixation has little effect on ultrastructure.

Collidine (sym-collidine, 2,4,6-trimethyl-pyridine) was introduced by Bennet and Luft (1959) as another OsO_4 buffer. It may be used during post-fixation (Gil and Weibel, 1968), for example, when cacodylate-buffered aldehyde is used during primary fixation. However, it is toxic, is expensive, and does not show any advantage over phosphate buffers (Glauert, 1975).

Various authors have taken a completely different approach by using organic solvents and fluorocarbons in place of a buffer. Epling and Sjostrand (1962) obtained good preservation using 40% osmium in carbon tetrachloride (also see Hobbs, 1969). Thurston et al. (1976) used more moderate concentrations of OsO_4 in fluorocarbons, and Zalokar and Erk (1977) carried the osmium in organic solvents. Although all of these authors achieved preservation comparable to that of buffered osmium tetroxide solutions the traditional buffering methods are most popular.

The reactions between osmium tetroxide and various macromolecules have been summarized by Millonig and Marinozzi (1968), Hayat (1970), Riemersma (1970), and Rodriguez-Garcia and Stockert (1979). It is stressed that osmium may act as a fixative, a stain, or both. For example, it will fix and stain unsaturated lipids, but only stabilizes some proteins. These factors are distinguished on the basis of the degree of osmium uptake by the specific type of macromolecules (Schiechl, 1971; Collins et al., 1974; Griffith et al., 1978; Nielson and Griffith, 1978).

Osmium tetroxide oxidizes unsaturated lipids (Criegee, 1936, 1938; Criegee et al., 1942) at the rate of one atom of osmium per lipid double bond (Riemersma and Bouijn, 1962; Stoeckinius and Mahr, 1965; Litman and Barrnett, 1972):

$$\underset{\text{Unsaturated fatty acid}}{-\overset{|}{C}=\overset{|}{C}-} + OsO_4 \longrightarrow \underset{\text{Cyclic osmic acid monoester}}{O_2Os\begin{matrix} O-\overset{|}{C}- \\ | \\ O-\underset{|}{C}- \end{matrix}} + 2H_2O \longrightarrow$$

$$\underset{\text{Osmic acid}}{H_2OsO_4} + \underset{\text{Diol}}{\begin{matrix} -\overset{|}{C}-OH \\ | \\ -\underset{|}{C}-OH \end{matrix}} + \left[O_2Os\begin{matrix} O-C \\ | \\ O-C \end{matrix} \right] \longrightarrow \underset{\text{Diester}}{\begin{matrix} C-O & & O-C \\ | & Os & | \\ C-O & & O-C \end{matrix}}$$

The final product of the reaction is a very stable diester that can resist dehydration and embedding (Ashworth et al., 1966). A great deal of research has been conducted with unsaturated lipids; the interested reader should consult the following for detailed analyses: Casley-Smith (1967), Riemersma (1968), Stein and Stein (1971), Collins et al. (1974), and White et al. (1976).

At the electron microscope level, unsaturated lipid droplets exhibit high contrast (Casley-Smith, 1967). Depending upon the size of the droplets, they may show an external-to-internal range of black-to-gray tones, mimicking penetration depths of large pieces of tissue. The droplets are not necessarily fixed in the exact position they occupied in the living cell because partial displacement and/or extraction is possible until the unsaturated lipid is stabilized (Stein and Stein, 1971). This also occurs in organelles; Fleischer et al. (1967) have shown that up to 80% of the total unsaturated lipid of mitochondria must be extracted before morphological integrity is lost, and even with 95% extraction cristae are recognizable. The addition of calcium to buffered OsO_4 will decrease lipid extraction (Strauss and Arabian, 1969). In short, using the general fixation method discussed here, one cannot quantitatively describe the unsaturated lipid content of a cell simply on the basis of stain patterns (Morgan and Huber, 1967).

This also indirectly indicates that the unsaturated lipid fraction of a tissue greatly influences osmium uptake. For example, tissues rich in unsaturated lipids (e.g., brain, pancreas, and kidney) rapidly react with the OsO_4, but the rate rapidly falls off with time as a barrier is formed; in contrast, tissues having a low lipid content (e.g., muscle and skin) only accommodate a fraction of the osmium when the fixation time is identical (Bahr, 1955). Having some idea of a tissue's unsaturated lipid content will assist one in micrograph interpretation, but, in general, it is unnecessary to accommodate these differences by increasing fixation time, although tissue cube size should be monitored.

Saturated lipids react with osmium, but the osmium is typically displaced or oxidized during dehydration (Schidlovsky, 1965; Hayat, 1970). Chapman and Fluck (1966) noted that stabilization will occur if fixation proceeds at 60°C, but for obvious reasons fixation is not conducted at high temperature. Consequently, saturated lipids are typically not retained during conventional fixation and embedding.

Problems exist in evaluating the role of osmium tetroxide in protein preservation; modifications induced by primary fixation in protein molecules drastically alter the protein, and subsequent reaction with OsO_4 is therefore with cross-linked carbonyl compounds. Although research has been conducted on pure protein suspensions, conflicting data exist; only general statements can be made concerning the role of protein preservation by OsO_4. Hopwood (1969a, b), Riemersma (1970), and Nielson and Griffith (1979) contain more comprehensive information.

Proteins containing tryptophan show a higher degree of reactivity toward osmium than those without it (Porter and Kallman, 1953;

Millionig and Marinozzi, 1968), as do sulfur-containing amino acids (Bahr, 1954; Adams, 1960; Hake, 1965; Elleder and Lojda, 1968a,b). Hopwood (1969b) also indicates that osmium, although in trace amounts, will react with phenolic, hydroxyl, carboxyl, amino, and heterocyclic groups. On the other hand, Adams and Bayliss (1968; Adams et al., 1967) contend that osmium reacts with protein-bound lipid. This matter was simplified by Elleder and Lojda (1968b), who concluded that proteins are indeed somewhat stabilized by osmium, but the degree of the reaction is insufficient to enhance electron contrast. Low concentrations of tissue-bound osmium may be detected during the OTAN reaction (osmium tetroxide-α-naphthylamine); thus, Elleder and Lojda (1968a) demonstrated that osmium is present in a complex form after protein osmication.

To study the reaction between lipoproteins and osmium, Dreher et al. (1967) used lung lipoprotein myelenics and observed the reactive sites by EM. They concluded that the lipid component was more reactive toward the OsO_4 than the protein. Concurrently they drew some conclusions concerning trilaminar (lipid–protein–lipid) membranes: the hydrophobic lipid layers can double in thickness while the hydrophilic protein layer remains dimensionally unaffected by osmication. This increase in membrane thickness is a function of the extent of the reaction, not necessarily the volume changes on the entire cell induced by osmium (Pentilla et al., 1974; Mersey and McCully, 1978); the permeability properties of the native membrane are drastically changed (Tormey, 1965; Jost et al., 1973)—that is, the entire cell is affected. In addition to the volume changes noted above, Millonig and Marinozzi (1968) observed that osmium fixation may cause swelling in isolated cells, and at the ultrastructural level Moretz et al. (1969b) noted myelin shrinkage. Other researchers have shown that among organisms, the appearance of cells is a function of sex and genetic background (Simson et al., 1978), age (Lawton and Harris, 1978), and the prevailing metabolic state (Gale, 1977).

Osmium has no effect on purified DNA or RNA (Bahr, 1954), and different carbohydrates show little or no stain or fixative effect following osmication (Millonig and Marinozzi, 1968). Although most carbohydrates are extracted during fixation and embedding, glycogen persists, probably because of its insolubility, or as the result of aldehyde and some OsO_4 modification (Millonig and Marinozzi, 1968;

DeBruijn, 1973). On the other hand, nucleic acid denaturation should be avoided; nucleic acids are fixed and stained with uranyl acetate (Hayat, 1968). In summary the following conditions are met for osmium tetroxide post-fixation:

Temperature: 0–4°C
Tissue size: ⩽0.5 mm^3
Time: 1–2 hr
Concentration: 2%
Buffer: phosphate or cacodylate
pH: 7.2–7.4
Final osmolarity: ∿0.1 M

Osmium tetroxide is an extremely hazardous compound that must be carefully handled to avoid self-exposure. Sax (1975) warns that both acute and chronic exposures to osmium causes ocular disturbances, contact dermatitis at low concentrations and ulceration at higher concentrations, and bronchitis. If its chlorinelike odor is detected, the lab should be cleared and throughly aired before personnel are readmitted; the threshold limit of OsO_4 is 0.002 mg/m^3. Therefore, open containers of osmium are always handled in a fume hood, and the user should also wear gloves and goggles.

During osmication of specimens, they should be held in a glass-stoppered vial because cork or rubber stoppers and parafilm will also react with the osmium. Osmium vapors will also blacken the interior of a refrigerator; the best way to avoid this but still fix at 0–4°C is to place the sealed vial in an ice-cooled water bath held within a fume hood. About halfway through fixation it may be necessary to replenish the ice bath.

Osmium tetroxide may be purchased in the crystalline form (1 g/vial) or in aqueous solution (usually 4%; 5–10 ml/vial). A major disadvantage of purchasing solid OsO_4 is that it requires a considerable length of time to dissolve in the buffer (∿24 hr), whereas the aqueous solutions may be simply diluted to the appropriate concentration with buffer just prior to use. This avoids the problem of storing stock solutions; the glass ampoules of osmium are refrigerated and boxed to avoid breakage. When buffered solutions are kept on hand, they are stored in amber, glass-stoppered vessels, and the entire bottle is wrapped with foil. Hayat (1970) recommends cold storage, but Glauert (1975) indicates that storage at room temperature will not harm the solution. Used solutions are recoverable and may be repeatedly used; see Schlatter and Schlatter-Lanz (1971) and Kiernan (1978) for simple methods.

The glassware used to prepare and contain the fixative must be extremely clean. Hayat (1970) recommends the following procedure:

1. Clean the glassware with soap and water, followed by concentrated nitric acid to remove organic contaminants.
2. Rinse several times with distilled water, and air-dry the glassware.
3. Because osmium tetroxide rapidly volatilizes, even under strict conditions, prepare the final fixative just prior to use.
4. Place the ampoule of OsO_4 in the cleaned storage bottle and break its seal with a cleaned glass rod. Work in a fume hood.
5. Add the appropriate buffer for dilution, mix, and initiate fixation.
6. Label the stock solutions with appropriate warnings.

Osmication is followed by washing with cold distilled water or buffer, which serves to remove excess OsO_4 from the tissue (cf. Ockleford, 1975). This avoids nonspecific OsO_4 reduction during dehydration, which can seriously embrittle the tissue and render it unsuitable for thin sectioning. The duration of washing must be short to prevent extraction or swelling, but it should be thorough. Thus, the author recommends several changes of the washing fluid and agitation over a 15–30-min. time period.

GLUTARALDEHYDE-OSMIUM TETROXIDE FIXATIVE

Various degrees of success have been achieved when primary and post-fixation are simultaneously carried out using a mixture of glutaraldehyde and osmium tetroxide (Glauert, 1975). This method was first proposed to avoid the artifacts inherent in double fixation and the buffer wash (Trump and Bulger, 1966). The researchers also noted that fixation must be carried out at 0–4°C, because the two fixatives react with one another at higher temperatures; fixation time must not exceed 1 hr, because the same reaction will become predominant and cells will not be well preserved (Hirsch and Fedorko, 1968; Franke et al. 1969).

Hopwood (1970b) indicated another problem, in that the individual fixative molecules may negatively compete with one another during reaction; for example, proteins may be blocked by osmium instead of the preferred glutaraldehyde cross-linking. Until more is learned about these reactions, the traditional primary followed by post-fixation method is recommended.

URANYL ACETATE

The potential of uranyl acetate as a nucleic acid fixative was first explored by Kellenberger et al. (1958), Ryter and Kellenberger (1958a), and Huxley and Zubay (1961) in their studies of bacteria. Hayat (1968, 1969) and Mumaw and Munger (1971) showed the efficiency of nuclear stabilization by uranyl acetate following double fixation, and recommends its general use as a tertiary fixative. It is effective as both a fixative and a stain for phospholipids (Silva et al., 1968, 1971; Hayat, 1970; Silva, 1973; Ting-Beall, 1980). Extraction of previously unfixed macromolecules, however, is likely (Farguhar and Palade, 1965). Nonetheless, application of uranyl acetate en bloc increases tissue contrast, and this effect is further enhanced by post-staining thin sections with uranyl acetate. (Lombardi et al., 1971). Hayat (1975) thoroughly covers the chemistry of the reactions between uranyl acetate and macromolecules. Below are the more general applications of uranyl acetate.

Uranyl acetate may be applied as a saturated (∿1.5%) aqueous solution immediately following the specimen's water wash (de Petris, 1965; Terzakis, 1968), or in the first dilute step of dehydration (e.g., 2% uranyl acetate in acetone). When fixatives are prepared with phosphate or cacodylate buffers, thorough water washing must precede uranyl acetate treatment to avoid precipitation (Farguhar and Palade, 1965).

Terzakis (1968) and Hayat (1970) indicate that aqueous uranyl acetate enhances contrast better than that used in dehydration; pH affects the ionic stability of the uranyl acetate molecule. At low pH (4–6) the molecule is most stable and preferentially reacts with DNA (Zobel and Beer, 1961; Wolfe et al., 1962, Séchaud and Kellenberger, 1972), and at higher pH (~5.0) the metal exists in a complex form and functions as a general stain. In short, uranyl acetate enhances the staining of osmicated tissues, and subsequent post-staining with heavy metal salts leads to excellent overall contrast.

Table 7-4. En bloc stains.

STAINING METAL	REACTIVE SITE	REFERENCE
Bismuth	Nucleic acids	Albersheim and Killias, 1963
Chromium	Cell membranes	Bullivant and Hotchin, 1960
Indium	Nucleic acids	Watson and Aldridge, 1961
Iron	Acid mucopolysaccharides	Albersheim et al., 1960
Ruthenium red	Acid mucopolysaccharides	Luft, 1966
Vanadium	Glycogen	Callahan and Horner, 1964

Stock solutions of uranyl acetate rapidly degrade with time; if the solution has a precipitate or is cloudy, discard it. Stock solutions are stable for about 1 week if stored in tightly sealed amber bottles. The author has found it useful to freeze small vials (∿2 ml) of aqueous uranyl acetate, and remove what is needed; the same solutions may also be used for post-staining. Uranium is a hazardous compound and as such must be carefully handled (Darley and Ezoe, 1976).

A summary of the conditions for tertiary fixation/en bloc staining with uranyl acetate is as follows:

Temperature: 0–4°C
Time: 15 min.
Concentration: 1.5–2%

Hayat (1970, 1975) discusses a large variety of heavy metal stains used for site-selective reactions. For example, lanthanum may be used to infiltrate the T-tubules of muscle without affecting other tissue components (Forbes and Sperelakis, 1971). Table 7-4 lists some site-selective stains that may be applied en bloc; Hayat (1975) and references therein should be consulted for specific methods. Not all heavy metals may be used as en bloc stains, because they may embrittle the tissue to a point at which thin sectioning is very difficult, or extraction of the stain occurs during dehydration and embedding.

DEHYDRATION

The majority of embedding media used for electron microscopy are immiscible with water; thus tissue fluids must be replaced with another fluid that will mix with water and the monomer (liquid) form of the embedding medium. This substitution is referred to as dehydration and typically employs a graded series of acetone or ethanol. Unfortunately, these are excellent solvents that will extract unfixed macromolecules (especially lipids), but sufficient data are available that artifacts induced by dehydration may be interpreted. When special experimental conditions are being enforced by the microscopist, the preparation steps should be studied ahead of time to obtain optimal results. For example, if selective staining using potassium permanganate is desired, embedding media containing nadic methyl anhydride are unsuitable; a reaction between the two occurs (Reedy, 1965). In the example given, the choice of a different embedding medium may require the use of a different reagent for dehydration. Good reviews of dehydration are available in Kahn et al. (1977) and Boyde and Maconnachie (1976); Howard and Postek (1979) have published a bibliography on dehydration that deals more with physical dehydration, but is still fairly complete for organic dehydration. Also recall the addition of uranyl acetate to low concentrations of the dehydration reagent for en bloc staining/fixation.

Regardless of which reagent is employed, a number of conditions are inherent in dehydration. First, unfixed macromolecules, especially lipids, are extracted (Ashworth et al., 1966), but the degree of extraction is a function of the specific reagent. Second, the application of a graded series of the reagent in water, as opposed to the application of an absolute concentration, decreases distortion: Cohen (1979) has noted that the effects of a graded series of 30%, 50%, 70%, 85%, 95%, 100%, and 100% reagent, as opposed to 70%, 100%, and 100%, are optimal. The recommended length of time at each step is 5–10 min. for 30% and 50%, with an increase to 10–20 min. for each of the remaining steps. The final reagent concentration is always repeated twice to ensure thorough dehydration, which in turn permits good infiltration of the embedding monomer, and finally results in a specimen suitable for thin sectioning. To reduce the degree of extraction and distortion, which always increases with time (Boyde and Maconnachie, 1979) and is an important factor with plant cells, dehydration may be accelerated by (Tormey, 1965; Coulter, 1967; Benscome and Tsutsume, 1970):

1. Conducting these steps at room temperature rather than at 0–4°C. (Hayat, 1970, conventionally recommends 0–4°C, with warming to room temperature at the final 100% reagent.)

2. Constantly agitating the specimens.
3. Frequently changing (every few minutes) the solution of the same concentration.
4. When dehydrating with ethanol for Epon embedding, infiltrating after 70% ethanol with a graded series of ethanol: Epon leading up to 100% Epon (Idelman, 1964, 1965). Thus, the shrinkage implicity associated with ethanol at concentrations ⩾70% is avoided.

Any or all of these factors are useful; in the author's laboratory, steps 1 through 3 are routinely applied for all tissue types, with maximum adherence when plant tissues are being processed. The reader must understand that complete dehydration is an essential part of fixation and embedding because it is prerequisite for total infiltration of the embedding monomer, which affects sectioning. Poorly embedded tissues simply cannot be thin-sectioned.

Acetone has been successfully used for a wide variety of specimen types, ranging from cultivated cells (e.g., Collins et al., 1980) to animal tissues (Nowell and Pawley, 1980). Initially, tissues exposed to low concentrations of acetone will swell, but they begin shrinking at 70–80% concentrations, where shrinkage over time is also at its maximum (Boyde et al., 1977; Boyde and Maconnachie, 1979). Lipid is extracted during dehydration, but the degree of extraction is debatable, probably according to specimen type: Ashworth et al. (1966) claim reduced extraction, while Cohen (1979) has observed that acetone extracted sufficient lipid to distort plant cell surfaces (using the SEM). On the other hand, Hayat (1970) wrote that acetone does not react with unreduced osmium tetroxide, it does not inhibit embedding monomer polymerization if trace amounts persist in the tissue, and it is relatively rapid. On the other hand, acetone characteristically embrittles tissue, and stock solutions are hydrophilic, meaning that dehydration may not be complete owing to water contamination.

In comparison, ethanol will not embrittle tissue, is less volatile, and is less toxic than acetone. Boyde and Maconnachie (1979) indicate that swelling occurs at low concentrations (⩽70%), followed by shrinkage at higher concentrations (maximum at 80–90%). Adams and Bayliss (1968) indicate that more phospholipid is extracted by ethanol than acetone, but, in general, less overall extraction occurs

(Millonig, 1964). Cohen (1979) notes that ethanol is less likely to dissolve unbound lipids, and Parsons et al. (1974) recommend that ethanol is somewhat better than acetone for plant tissues. A major disadvantage of ethanol is that mixtures of it and the monomer of epoxy resins slowly penetrate tissue cubes. Luft (1961) introduced an intermediate step with propylene oxide, which serves as a transition; that is, ethanol dehydration ⟶ ethanol: propylene oxide ⟶ 100% propylene oxide ⟶ propylene oxide: epoxy resin ⟶ pure epoxy resin. Propylene oxide can remain in the embedding medium, even during polymerization (Grossenbacher and Hardley, 1969), extracts lipid (Bushman and Taylor, 1968), and extracts phosphotungstic acid when used for en bloc staining (Glauert, 1975). Propylene oxide is also toxic and reactive; always work with it in a fume hood away from heat.

Although either acetone or ethanol may be used for general tissue preparation, the literature should be checked for special application. For example, Page and Huxley (1963) noted that striated muscle filaments are 10% shorter after dehydration in ethanol than they are in acetone. Similarily, Johnston and Roots (1967) observed that cell junctions in brain are better preserved with acetone than ethanol.

Two reagents have recently been introduced for dehydration, but few data have yet been published. Acidified 2,2-dimethoxypropane, or DMP, upon reaction with water yields methanol and acetone (Muller and Jacks, 1975). Although Kahn et al. (1977), Maser and Trimble (1977), and Thorpe and Harvey (1979) report results comparable to ethanol dehydration, Boyde et al. (1977) note distortion from shrinkage. DMP dehydration of plants was successful for Lin et al. (1977), who also obtained similar results with 2,2-diethoxypropane (which yields acetone and ethanol when hydrated).

EMBEDDING

The final stage of tissue preparation involves infiltrating the specimen with a material that supports the specimen without distortion but is still sufficiently elastic to be cut into sections 200–800 Å thick. Immediately following dehydration the tissue is treated by:

1. Infiltration with a mixture of the dehydration reagent and the embedding medium.

2. Infiltration with pure embedding medium.
3. Placing the tissue in a labeled embedding mold.
4. Polymerization (usually by heat) of the embedding medium.

This is the most time-consuming preparation step; under optimal conditions it requires ∿17 hr, although some rapid embedding methodologies may be adapted for individual needs (Hayat and Giaquinta, 1970; Shinagawa et al., 1980).

Hayat (1970) listed the characteristics of a suitable EM embedding medium: it should be inert toward the preserved and dehydrated tissue, its volume should not significantly change during polymerization, it must be radiation-resistant and stable during electron microscope observations, and finally it should be sufficiently plastic to support the tissue while also being elastic enough to permit thin sectioning. Unfortunately, no embedding medium available today fulfills all of these criteria, but enough is known to interpret the artifacts induced by embedding (e.g., fat extraction). A completely different approach is being investigated in several laboratories; cryòultramicrotomy involves sectioning of frozen, nonembedded tissues (e.g., Saubermann, 1980).

Basically, embedding media are either water-miscible or -immiscible. Whereas the former are typically used only for special applications (e.g., electron cytochemistry), embedding media insoluble in water are overwhelmingly popular. Although immiscible media require dehydration with, for example, acetone, they come very close to fulfilling all of the criteria noted above. In comparison, water-miscible infiltration and dehydration occur simultaneously, but comparatively more extraction and displacement accompany these steps. Table 7-5 compares the various water-insoluble media and water-miscible media.

Water-Immiscible Embedding Media

Methacrylates, epoxy resins, and polyester resins are representative of water-insoluble embedding media. The methacrylate embedding media, with the exception of water-soluble hydroxypropyl methacrylate, are only noted in passing because better media have been developed. Introduced by Newman et al. (1949), methyl methacrylate (an acrylic resin) cannot tolerate electron irradiation (it will flow during examination) and shrinks considerably (∿20% during

Table 7-5. Characteristics of embedding media.

	WATER-INSOLUBLE EMBEDDING MEDIA						WATER-MISCIBLE EMBEDDING MEDIA			
	EPOXY RESINS			POLYESTER RESINS		METHACRYLATES	EPOXY-BASED		METHACRYLATE-BASED	
	Poly Bed 812[a]	Araldite[b]	Spurr[c]	Vestopal[d]	Rigolac[e]	Methyl methacrylate[f]	Aquon[g]	Durcupan[h]	HPMA[i]	GMA[j]
Solubility in common dehydration reagents	Good			Good		Good	N/A			
Infiltration rate	Slow	Slow	Rapid	Slow		Rapid	Slow			
Polymerization rate	Good, heat			Good, heat, UV light		Good, heat or light	Slow, UV			
Deformation	Very low			Very low		20% shrinkage	Moderate			
Radiation resistance	High			High		Mass loss 50%	Good			
Sectioning qualities	Good			Difficult		Good	Moderate			
Pot life	Long, frozen			Rapid deterioration		Spontaneous polymerization	Short			
Toxicity	Moderate			Moderate		Low	High			

[a]Polysciences trademark based on Luft, J. H. (1961) *J. Biophys. Biochem. Cytol.* 9:409.
[b]Glauert, A. M., G. E. Rogers, R. H. Glauert (1956) *Nature* 178:803.
[c]Spurr, A. R. (1969) *J. Ultrastr. Res.* 26:31.
[d]Ryter, A., E. Kellenberger (1958b) *J. Ultrastr. Res.* 2:200.
[e]Kushida, H. (1960) *J. Electron Micros.* 9:113.
[f]Newman, S. B., E. Borysko, M. Swerdlow (1949) *Science* 110:66.
[g]Gibbons, I. R. (1959) *Nature (London)* 184:375.
[h]Staubli, W. (1960) *C. R. Searc. Soc. Biol.* 250:1137.
[i]Hydroxypropyl methacrylate. Leduc, E. H., S. J. Holt (1965) *J. Cell Biol.* 26:137.
[j]Glycol methacrylate. Rosenberg, M., P. Bartl, J. Lesko (1960). *J. Ultrastr. Res.* 4:298.

polymerization; Reimer, 1959). However, variations of this medium have proved very useful for light microscope histology, for preserving bulk samples (e.g., insects), and as a precursor of water-miscible media.

Epoxy Resins

The methacrylates were replaced in the 1950s by epoxy resins, which polymerize (by heat) uniformly with minimal shrinkage and are more radiation-resistant than methacrylates (Grund et al., 1978). The most widely used epoxy resins are Araldite (Glauert et al., 1956), Epon (Finck, 1960; Luft, 1961), and Spurr (Spurr, 1969). All epoxy resins characteristically contain an epoxy group, $\overset{O}{\overbrace{C-C}}$, at both ends of a molecule of polyaryl ethers of glycerol (Figure 7-2a). With the addition of a suitable hardener, for example, dodecenyl succinic anhydride (DDSA) or nadic methyl anhydride (NMA) (Figure 7-2b), the epoxy will cross-link in three dimensions and form a very stable polymer. The polymerization is catalyzed by the addition of an accelerator such as tridimethylaminomethylphenol (DMP-30, Figure 7-2c) and heat. Grund et al. (1978) and Buchanan and Stewart (1980) recently compared a number of these resins.

Araldite. Glauert and her coworkers (Glauert et al., 1956; Glauert and Glauert, 1958) were the first researchers to successfully formulate an epoxy resin for electron microscopy. In the United States, a modification of Glauert's basic formulation is marketed as Araldite 502 (Finck, 1960; Luft, 1961). Although it is readily miscible with acetone, if trace amounts of ethanol are retained by the tissue, it will interfere with polymerization, a problem that requires transition with propylene oxide (Luft, 1961). Luft's (1961) formulation is as follows:

Araldite 502	100 g
DDSA	75 g
DMP-30	2.3–3.5 g

The ratio of resin to hardener may be modified to produce a harder or softer block. Polymerization is complete within 24 hr at 60–70°C

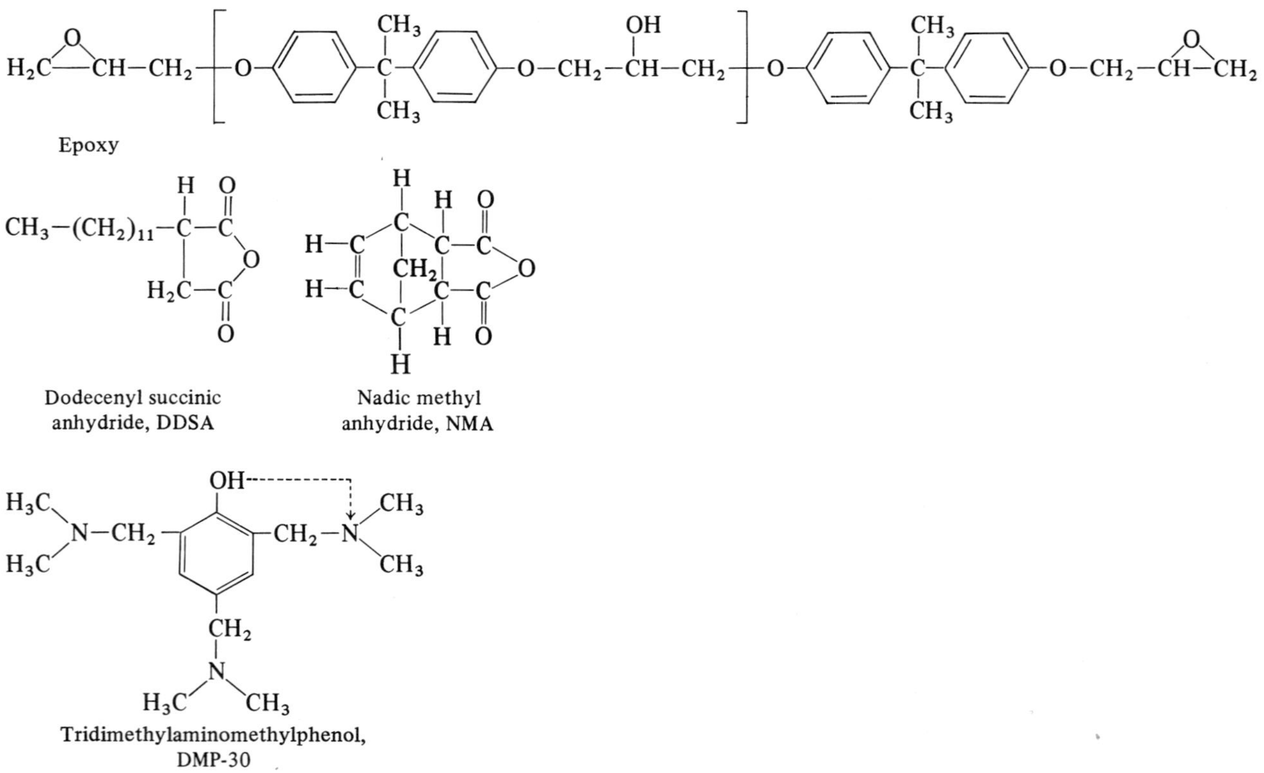

CH$_3$ OH CH$_3$

O O

H$_2$C—CH—CH$_2$—[O—C—O—CH$_2$—CH—CH$_2$]—O—C—O—CH$_2$—CH—CH$_2$

CH$_3$ CH$_3$

Epoxy

H O

CH$_3$—(CH$_2$)$_{11}$—C—C

O

H$_2$C—C

O

H

H O

C

H—C C—C

CH$_2$ O

H—C C—C

C

H O

H

Dodecenyl succinic anhydride, DDSA

Nadic methyl anhydride, NMA

OH

H$_3$C CH$_3$

N—CH$_2$ CH$_2$—N

H$_3$C CH$_3$

CH$_2$

N

H$_3$C CH$_3$

Tridimethylaminomethylphenol, DMP-30

Figure 7-2. Components of the epoxy resins Araldite and Epon-Poly/Bed.

although a more stable block will result if a schedule of overnight at 34°C, full day at 45°C, and finally another full day at 60°C is followed. The author has satisfactorily polymerized blocks by heating them for 1-2 hr at 70-80°C followed by overnight incubation at 60°C. Letting the blocks cool and sit for a few hours (4-24 hr) greatly assists thin sectioning.

Epon. Concurrent with the above work, Luft (1961) presented Epon 812 as an epoxy resin suitable for embedding. This was by far the most popular embedding medium for most applications, but the basic resin is no longer supplied by the Shell Oil Company, and consequently the traditional Epon 812 is not available. Polysciences Inc., however, currently markets a replacement, Poly/Bed 812, which is an adequate substitution, having all the qualities of Epon 812. Ernest F. Fullam, Inc. supplies Epok 812, again a good replacement.

A number of people contributed to the formulation of Epon-Poly/Bed (Kushida, 1959; Finck, 1960; Luft, 1961). Its primary advantage is that it is relatively less viscous than Araldite; in addition, the hardness of the final block is easily adjusted by changing the ratios of its anhydride hardeners, DDSA and NMA. This advantage is useful when working with dense tissues such as bone. The Polysciences formulation is as follows:

Poly/Bed 812	21 ml
DDSA	13 ml
NMA	11 ml
DMP-30	0.7 ml

When ethanol is employed for dehydration, a transitional step with propylene oxide is recommended; acetone is readily miscible with the final medium. Polymerization is complete overnight at 60°C, or using this schedule: overnight at 35°C, next day at 45°C, next night at 60°C. As with Araldite, a day or so of curing at room temperature improves sectioning.

Mollenhauer (1964) proposed a mixture of Araldite and Epon-Poly/Bed that has very good sectioning properties and offsets the disadvantages of both media. His formulation, modified by Polysciences, is as follows:

Poly/Bed 812	100 ml
Araldite 502	60 ml
DDSA	180 ml
DMP-30	5.2 ml

Polymerization is the same as described above.

Araldite and Epon/Poly-Bed, whether used alone or as a mixture, must be thoroughly stirred for a homogeneous medium to result. If incorrect ratios of the components, incomplete mixing of the final medium, or insufficient infiltration of the tissue occurs, thin sectioning is impossible. Because standard formulations are very well known, they must be used despite their being tedious.

Wherever possible, gravimetric rather than volumetric methods should be used; careful weighing of each successive component into the same vessel avoids loss by transfer into another vessel, and ratios among components are easier to detect. If glassware is used, it should be immediately cleaned with acetone because these polymers are insoluble in other conventional solvents. Alternatively, disposable polyethylene containers may be used. After the components are combined, the mixture must be thoroughly stirred (~20 min.) for complete homogeneity; avoid introducing air bubbles.

Stock solutions of epoxy resins not containing the accelerator may be frozen away from water for several months, without an increase in viscosity. Before use they should be brought to room temperature and the accelerator added.

The method of infiltration for acetone-dehydrated specimens using Araldite, Epon-Epoxy/Bed, or a mixture is as follows:

1. When the specimens are in acetone, prepare the final embedding medium; that is, add the accelerator to the stock solution. Prepare a 1:1 infiltration fluid by mixing equal volumes of acetone and final medium.
2. Decant the final 100% acetone change, and replace it with the acetone-medium for 30 min. Agitate frequency to maintain homogeneity of the infiltration mixture. Note: Dense specimens may require several dilutions of the acetone-medium (e.g., 35, 50, 75%) if infiltration has been determined to be incomplete in previous preparations.

3. Decant the mixture, and replace it with 100% embedding medium for 30 min. Agitate to remove acetone. A very mild vacuum by aspiration may be used for dense specimens, but it is not recommended unless very carefully controlled.
4. Embed in predried capsules or molds according to polymerization schedules given above. The method of embedding is discussed below under "Comments on Resin Embedding."

When ethanol is used for dehydration, propylene oxide transition is necessary; substitute propylene oxide for acetone in the above method. A variety of mechanical shakers are available for assisting infiltration; see Jurand and Ireland (1965), Steinbrecht and Ernst (1967), Kushida (1969), Banfield (1976), Goldfarb et al. (1977), and Sjostrand and Halma (1978).

Spurr. A major disadvantage of Epon-Poly/Bed and Araldite is their high viscosity at room temperature (3000 cps and 150–210 cps, respectively). Consequently, infiltration of either medium into dense tissues is lengthy. Spurr (1969) solved this problem by introducing another epoxy resin, Spurr Low-Viscosity Embedding Medium; freshly prepared formulations have a viscosity of 60 cps, meaning that infiltration of dense tissues (especially plants) is rapid and thorough. The components of Spurr are quite different from those of the other epoxies (Figure 7-3). The basic resin is vinylcyclohexene dioxide (VCD), also known as ERL-4206, and another resin functioning as a flexibilizer is diglycidyl ether of polypropylene glycol (DER-736). It also contains the hardener nonenyl succinic anhydride (NSA) and the accelerator dimethylaminoethanol (DMAE). Polymerization is for 8 hr at 70°C. Spurr (1969) recommends the following formulations (rapid cure is 3 hr at 70°C):

Polymerized Block Characteristics

	Standard firm	Hard	Soft	Rapid
VCD	10.0 g	10.0 g	10.0 g	10.0 g
DER-736	6.0	4.0	7.0	6.0
NSA	26.0	26.0	26.0	26.0
DMAE	0.4	0.4	0.4	1.0

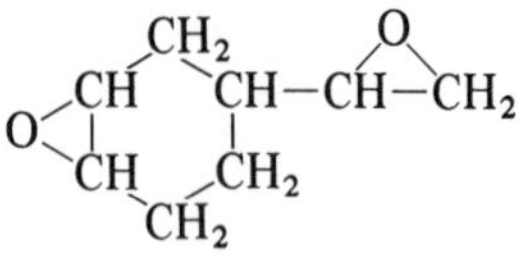

VCD = vinylcyclohexene dioxide

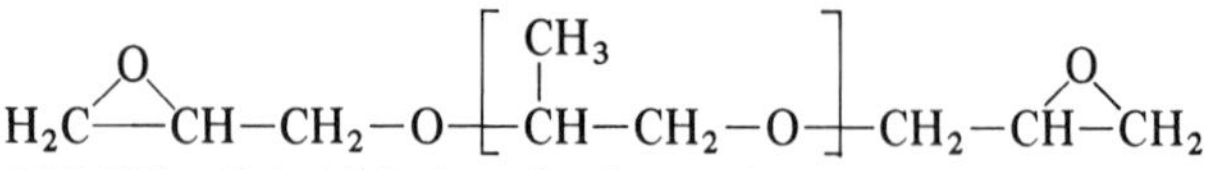

$$H_2C\overset{O}{-}CH-CH_2-O-\left[\underset{}{\overset{CH_3}{\overset{|}{C}H}}-CH_2-O\right]-CH_2-\overset{O}{CH-CH_2}$$

DER-736 = diglycidyl ether of polypropylene glycol

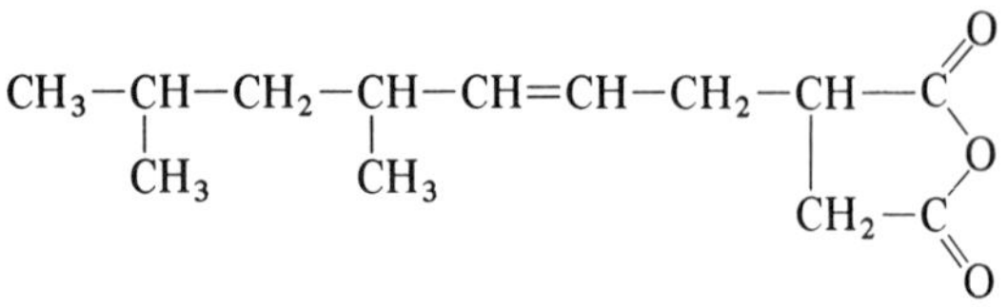

NSA = nonenyl succinic anhydride

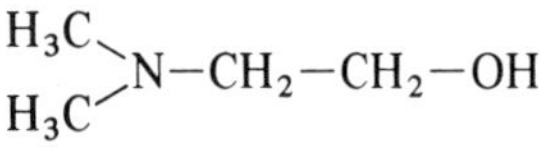

$$(H_3C)_2N-CH_2-CH_2-OH$$

DMAE = dimethylaminoethanol

Figure 7-3. Components of Spurr embedding medium.

Because this medium has very low viscosity, the individual components easily and readily mix, and its preparation is considerably more rapid than preparation of the other epoxies. Spurr (1969) recommends gravimetric preparation of this medium (i.e., the constitutents are consecutively added by weight to the same receptable). Anhydrous freezing of stock solutions without the accelerator gives preparations that are stable for several months.

The infiltration of low-viscosity Spurr into dense specimens follows a route different from that of the other epoxy resins. Following dehydration of the specimen in either 100% acetone or 100% ethanol (a transitional fluid is not required), an equal volume (relative to the volume of acetone or ethanol) of pure embedding medium is added to the specimen vial, swirled for thorough mixing with the dehydration reagent, and allowed to remain at room temperature for 30 min. Another equal volume of Spurr is added, swirled for homogeneity, and allowed to remain for 30 min. Decant this mixture and resuspend the samples in 100% Spurr for 30 min., or until infiltration is complete. Periodic agitation by hand or continual agitation using a rotary shaker (e.g., Kushida, 1969) will ensure that trace amounts of the dehydration reagent are displaced. The specimens are individually placed in oven-dried capsules and polymerized.

Polyester Resins

The introduction of polyester resins was concurrent with that of epoxy resins, their purpose also being to replace methacrylates. Vestopal W was introducted by Ryter and Kellenberger (1958b), and Rigolac by Kushida (1960). Vestopal W is commercially available in the United States and elsewhere, but Rigolac is difficult to obtain. Although the epoxy resins are more universally popular than polyester resins, the latter have the advantage that under certain conditions polymerization can be conducted under UV light rather than heat. This is important when thermally sensitive materials are prepared. On the other hand, the individual stock components have a short shelf life, the complete stock mixture is prone to contamination, the media are immiscible with ethanol, and finally two of the components (benzoyl peroxide and cobalt naphthenate) may explode if mixed together (Glauert, 1975). Consequently, these resins will not be dwelled upon; the reader should refer to the individual references for more complete information.

Vestopal W is composed of an unsaturated polyester resin (trade name Vestopal W), benzoyl peroxide (catalyst), and cobalt naphthenate (accelerator). Volumetric preparation of the final medium requires thorough mixing of 100 parts Vestopal W with 1 part of benzoyl peroxide; 0.5 part of the accelerator is then added and mixed. This stock medium is stable for several months when refrigerated in the anhydrous state. However, because polymerization is initiated as soon as the three components are mixed together, and polymerization is a function of time, temperature, and photoreaction, freshly prepared media are preferable. The recommended dehydration and infiltration procedure using acetone is as follows (Ryter and Kellenberger, 1958b): Infiltrate with 3:1, 1:1, 1:3 parts of acetone: Vestopal W for 30–60 min. at each step, followed by infiltration with the pure medium for several hours. Heat polymerization is completed within 12–24 hr at 60°C. If benzoin is used as a substitute for the normal catalyst (benzoyl peroxide), UV polymerization at room temperature for ~10 hr is possible (Kushida, 1964b).

Comments on Resin Embedding

Regardless of the resin chosen for embedding, a number of general rules must be followed for good results. First and foremost, preparation of stock solutions (whether by volumetric or gravimetric means) must be consistent and exact (Burke and Geiselman, 1971; Chang, 1971; Geiselman and Burke, 1973). Because the individual components are viscous (aggravating to measure exactly) but relative proportions determine sectioning characteristics, the initial volumes or weights must be carefully attended to. The most efficient polymerization will occur only if the medium is homogeneous; stirring with a glass or Teflon rod for ~20 min. is necessary. Magnetic stirrers are effective but introduce air bubbles into the medium. If magnetic stirrers are used, or when hand-stirring produces bubbles, the medium should be mildly vacuum-aspirated to remove the bubbles. Disposable plasticware should be used for preparing the media. Direct contact with the media results in contact dermatitis; gloves always should be worn during preparation, and working in a fume hood is an asset. Gorycki (1978) came up with an ingenious idea for mixing media: the components are placed in a plastic bag that locks shut, and the resin is mixed by rolling a metal rod over the surface. This thoroughly and rapidly mixes the resin without bubbles.

Stock solutions should be kept anhydrously frozen to minimize polymerization. In fact, low humidity during infiltration and embedding improves polymerization (Dellmann and Pearson, 1977). When stock solution viscosity increases, the solution should be discarded. Although in previous sections it has been recommended that the media be stored without the accelerator, the author has found it convenient when using Araldite, Epon-Poly/Bed, or a mixture of the two, to freeze the media containing the accelerator in disposable polyethylene syringes (~15 cc) for up to 10 days. Undesirable polymerization is minimal. The researcher need not determine the amount of accelerator required, and sufficient medium is available for preparing five specimens. After bringing the syringe to room temperature, dilute a small portion of its volume with the dehydration reagent; this avoids the question of: Did I infiltrate with the complete embedding medium? Infiltration with the properly formulated, complete medium is essential. The syringes are also very useful for filling the small embedding capsules or molds. Several smaller syringes (2–5 cc) may be filled with the monomer; these may be kept frozen and then used to coverslip thick sections for light microscopy. Because the tissue sections and coverslip medium are from the same batch, refractive index differences are avoided in light microscopy. The methods for embedding and molds are considered below under "Methods of Embedding."

Water-Miscible Embedding Media

Concurrent with the development of epoxy and polyester resins in the early 1960s, researchers explored the possibilities of using water-soluble embedding media, thus avoiding the rigors of organic dehydration and infiltration. Following conventional double fixation, specimens are infiltrated/dehydrated simultaneously with a graded series of distilled water and water-soluble media, which are then polymerized by exposure to ultraviolet light or heat. Unfortunately, in comparison to epoxy resins, relatively more distortion and/or extraction is readily apparent with water-miscible media. Consequently, they usually are restricted to studies involving electron histochemistry where the high temperatures required for immiscible media may deactivate enzyme systems (e.g., Ashford et al., 1972; Spaur and Moriarty, 1972). A few of the more common media are discussed below; Hayat (1970) deals much more fully with this topic.

Water-miscible embedding media are classified as derived from methacrylates or derived from epoxy resins. For example, Aquon and Durcupan are epoxy-based, whereas glycol methacrylate and hydroxypropyl methacrylate are based on acrylic resins. These are traditional media; other water-soluble embedments not derived from resins include gelatin (Bernhard and Nancy, 1964; Moretz et al., 1969a,b) and urea-formaldehyde (Casley-Smith, 1967; Peterson and Pease, 1970a,b). A completely different approach involving sectioning of unembedded tissues was introduced by Farrant and McLean (1969), who cross-linked albumin and glutaraldehyde; Pease (1980) recently modified this approach by mixing serum albumin with latex, and cross-linking with aldehyde. He has also altered urea embedding, using glutaraldehyde for cross-linking (Pease, 1973; Pease and Peterson, 1972; Peterson and Pease, 1972). Heckman and Barrnett (1973) cross-linked glutaraldehyde with carbohydrazide and produced GACH embedding medium. This, and the preceding nontraditional water-miscible media, are used when lipid retention is desired.

Aquon, which is the water-miscible fraction of Epon 812, was introduced by Gibbons (1958, 1959). Because it is not commercially available, the researcher must separate this phase from the resin (see Chapter 11). After isolation of the Aquon, the following stock solution is prepared and may be kept anhydrously frozen for several months:

Aquon	10 ml
DDSA	25 ml

The final embedding medium is prepared by the addition of the accelerator benzyldimethylamine (BDMA):

Stock Aquon	10 ml
BDMA	0.1 ml

Because this medium is more miscible with water at temperatures lower than 15°C, graded infiltration with Aquon/distilled water is done at 15°C. Infiltration with pure medium requires around 4 hr; polymerization is for 4 days at 60°C.

Shinagawa et al. (1980) introduced another Epon-based technique that uses semihydrated medium. It is stressed that this is not the conventional water-miscible technique because specimens are dehydrated in ethanol up to 90% concentration, followed by transition with propylene oxide, and finally with pure medium. The medium is prepared as follows:

Epon	6.0 ml
Distilled water	0.48 ml
Epon 815	4.0 ml
DMP-30	0.9 ml

Polymerization is at 55°C for 4 hr, or at 60°C for 2 hr. The advantages of this method are that dehydration and polymerization time are significantly reduced relative to conventional Epon or Aquon schedules.

Another water-miscible, epoxy-based resin is Durcupan, an aliphatic polyepoxide introduced by Staubli (1960). It can be purchased directly from various suppliers and is prepared as follows:

Durcupan	5 ml
DDSA	11.5 ml
Accelerator 960	1-1.2 ml
Dibutyl phthalate	0.2-0.5 ml

Thin sections of this medium normally exhibit a defect known as chatter; Kushida (1964a) thus offers another Durcupan formulation that results in a harder block. Using the above formulation, infiltration follows this schedule (Staubli, 1960): 50%, 70%, and 90% Durcupan in water for 30 min. at each step is followed by two 100% changes in pure Durcupan for 60 min. each. Polymerization is at 45°C for 24 hr. Severe dermatitis will result if the user handles this medium directly.

Kushida (1963, 1964a, 1965, 1966) has thoroughly investigated substitution of Durcupan for acetone or ethanol dehydration, followed by embedding in epoxy or polyester resins. No real advantage over conventional dehydration reagents was demonstrated, although an application may be found when acetone- or ethanol-soluble materials that are insoluble in Durcupan are studied.

Glycol methacrylate (GMA) is an acrylic-resin-based medium that was introduced by Rosenberg et al. (1960) and Leduc et al. (1963). Chemically it is 2-hydroxyethyl methacrylate and is miscible with water or ethanol. Hayat (1970) notes that the degree of preservation varies within embedded tissue blocks, membranes are in negative image, and GMA induces severe contact dermatitis; consequently, it is not recommended for general use. Leduc and Bernhard (1967) modified their earlier method in an effort to reduce swelling artifacts; their preparation of the stock GMA is as follows:

97% GMA + 3% distilled water	7 parts
98% butylemethacrylate + 2% Luperco	3 parts

The above mixture is prepolymerized for infiltration using this method:

1. A small portion of the medium is heated and continuously swirled until boiling just begins. Avoid water contamination.
2. While still swirling the mixture, plunge it into an ice bath until the solution temperature is ~2°C.

3. Steps 1 and 2 are repeated until the viscosity of the GMA at 0–4°C is similar to that of thick syrup.

The tissues are infiltrated at 0–4°C in 80% and 97% GMA/distilled water for 20 min. each, followed by 100% GMA for ~12 hr. The specimens are embedded in gelatin capsules (not polyethylene); polymerization will not occur in the presence of air. Polymerization with UV light (wavelength >3150 Å) is complete within 24–48 hr. Thin sections must be mounted on carbon or plastic-supported grids.

In an effort to overcome the problems associated with GMA, Leduc and Holt continued their research on water-soluble components of acrylic resins and introduced hydroxypropyl methacrylate, HPMA (Leduc and Holt, 1965). It is more readily miscible with water especially below 20°C and has better sectioning properties than GMA; on the other hand, penetration is significantly slower, as seen in this schedule: two changes of 85% and two changes of 97% HPMA in water are conducted for 1 hr, followed by infiltration with prepolymerized 100% HPMA for 1 hr. These steps must be carried out in a cold room and using a constant-agitation mechanical shaker. Polymerization by UV light at 10°C is completed within 12–24 hr; Alternatively heat polymerization (56°C, 2–3 days) may be used.

Methods of Embedding

Depending upon the application, different types of embedding forms are available. Polyethylene capsules, available commercially as BEEM embedding capsules (size 00), are used for embedding any type of routine tissue specimens (Figure 7-4a). They are preshaped into a square-topped pyramid, which greatly reduces the amount of post-polymerization trimming of the specimen block for sectioning. Before the advent of these molds, gelatin capsules with rounded ends were used for embedding, but they required extensive trimming for sectioning. Gelatin capsules are still used when high-temperature polymerization is conducted (e.g., for rapid tissue processing, polymerization is at 99°C); BEEM capsules will melt above ~80°C. Whether gelatin or polyethylene capsules are used, they should be predried in an oven at moderate temperature for 1–2 hr (gelatin becomes more brittle over time). The capsules are held for polymerization in com-

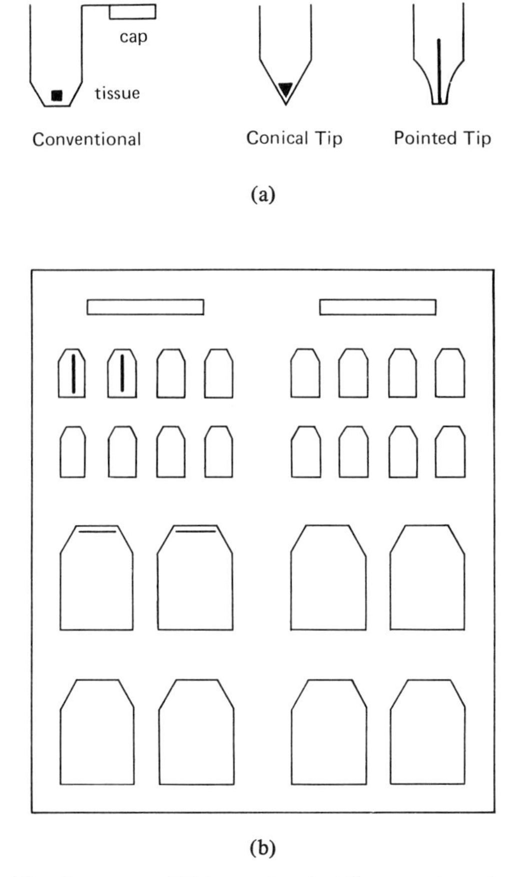

Figure 7-4. Embedding forms. a. BEEM capsules. b. Silicone or latex flat embedding mold.

mercially available block holders, or wooden blocks may be drilled to accommodate the capsules.

Different shapes of embedding molds are available for special applications. In addition to the shape discussed above, conical and bottle-neck capsules are available for, respectively, very small and long specimens (Figure 7-4a). Flat embedding molds (Figure 7-4b), in a variety of configurations, are available for orientation-embedding of a specimen (e.g., Chien, 1980). For example, a root hair may be embedded longitudinally for cross-sectioning (van der Wal and Dohmen, 1978). Whereas BEEM capsules are inexpensive and disposable, flat molds composed of silicon rubber are reusable, as flexing the mold releases polymerized specimens. Specimens embedded in polyethylene capsules readily pop out if loosened and then squeezed out using pliers; razor blades for cutting the capsule are not recommended.

Because all embedded tissues are similar in appearance, each and every capsule should be labeled. Small labels showing the date, specimen type, and any other pertinent information should be typed or penciled (do not use ink!) on paper and placed in each mold. This avoids considerable confusion when specimens accumulate; the label will be "embedded" in the polymerized embedding medium and thus will remain in the specimen block even after its removal from the mold.

The method used for embedding is as follows:

1. Label the oven-dried molds or capsules.
2. A few minutes before the infiltration (100% embedding medium) is concluded, fill the capsules or mold approximately one-third full with complete embedding medium. Wear rubber gloves.
3. Remove individual specimens with a wooden applicator stick, and place them at the bottom of the capsule or oriented in the mold. Position the tissue as close to the center of the capsule as possible, but avoid mishandling the specimen.
4. Prepare as many capsules as needed, and fill capsules nearly to the top. Avoid introducing air bubbles into the medium; the use of a polyethylene syringe is very convenient. Polyester resins should be capped.
5. Polymerize according to the standard method for the medium employed.
6. The cured blocks are now ready for sectioning.

REFERENCES

Abrunhosa, R. (1972) Microperfusion fixation of embryos for ultrastructural studies. *J. Ultrastr. Res.* 41:176.

Adams, C. M. W. (1960) Osmium tetroxide and the Marchi method: Reactions with polar and nonpolar lipids, protein, and polysaccharide. *J. Histochem. Cytochem.* 8:262.

——, and O. B. Bayliss. (1968) Reappraisal of osmium tetroxide and OTAN histochemical reactions. *Histochemie* 16:162.

——, Y. H. Abdulla, and O. B. Bayliss. (1967) Osmium tetroxide as a histochemical and histological reagent. *Histochemie* 9:68.

Albersheim, P., and U. Killias. (1963) The use of bismuth as an electron stain for nucleic acids. *J. Cell Biol.* 17:93.

——, K. Muhlethaler, and A. Frey-Wyssling. (1960) Stained pectin as seen in the electron microscope. *J. Biophys. Biochem. Cytol.* 8:501.

Albin, T. B. (1962) Handling and toxicology. In: *Acrolein*, p. 234. Smith, C. W. (ed.). John Wiley and Sons, New York.

Alexa, G., D. Chisalita, and G. Chirita. (1971) Reaction of dialdehyde with functional groups in collagen. *Rev. Tech. Ind. Cuir.* 63:5.

Altmann, P. L., and D. S. Dittmer (eds.). (1973) *Biological Data Handbook*, 2nd ed. Federation of American Society for Experimental Biology, Bethesda, Md.

Arborgh, P., P. Bell, U. Brunk, and V. P. Collins (1976) The osmotic effect of glutaraldehyde during fixation. A transmission electron microscopy, scanning electron microscopy, and cytochemical study. *J. Ultrastr. Res.* 56:339.

Artvinli, S. (1975) Biochemical aspects of aldehyde fixation and a new formaldehyde fixative. *Histochem. J.* 7:435.

Ashford, A. E., W. G. Allaway, and M. E. McCully. (1972) Low temperature embedding in glycol methacrylate for enzme histochemistry in plant and animal tissues. *J. Histochem. Cycochem.* 20:986.

Ashworth, C. T., J. S. Leonard, E. H. Eigenbrodt, and F. J. Wrightsman. (1966) Hepatic intracellular osmiophilic droplets: Effects of lipid solvents during tissue preparation. *J. Cell Biol.* 31:301.

Aso, C., and Y. Aito. (1962) Studies of the polymerization of bifunctional monomers. II. Polymerization of glutaraldehyde. *Macromol. Chem.* 58:195.

Bahr, G. F. (1954) Osmium tetroxide and ruthenium tetroxide and their reactions with biologically important substances. *Exp. Cell Res.* 7:757.

—— (1955) Continued studies about the fixation with osmium tetroxide. *Exp. Cell Res.* 9:277.

Baker, J. R. (1965) The fine structure produced in cells by fixatives. *J. Roy. Micros. Soc.* 48:115.

—— and J. M. McCrae. (1966) The fine structure resulting from fixation by formaldehyde: The effects of concentration, duration and temperature. *J. Roy. Micros. Soc.* 58:391.

—— and H. Rosenkrantz. (1976) Volumetric instillation of fixatives and inert substances into mouse lungs. *Stain Technol.* 51:107.

Banfield, G. W. (1976) Automation in tissue processing. In: *Some Biological Techniques in Electron Microscopy*, p. 166. Parsons, D. F., (ed.). Academic Press, New York.

Barnes, C. D., and L. D. Etherington. (1973) *Drug Doses and Laboratory Animals.* Univ. of California Press, Berkeley.

Beesley, J. A. (1978) A new technique for preparing cell monolayers for electron microscopy. *Stain Technol.* 53:48.

Bennet, H. S., and J. H. Luft (1959) s-Collidine as a basis for buffering fixatives. *J. Biophys. Biochem. Cytol.* 6:113.

Benscome, S. A., and V. Tsutsumi (1970) A fast method for processing biologic material for electron microscopy. *Lab. Invest.* 23:447.

Bernhard, W. and M. T. Nancy (1964) Coupes à congélation ultrafines de tissu inclus dans la gélatine. *J. Microscopie* 3:579.

Bodian, D. (1970) An electron microscopic characterization of classes of synaptic vesicles by means of controlled aldehyde fixation. *J. Cell Biol.* 44:115.

Bone, Q., and E. J. Denton (1971) The osmotic effects of electron microscope fixatives. *J. Cell Biol.* 49:571.

—— and K. P. Ryan (1972) Osmolarity of osmium tetroxide and glutaraldehyde fixatives. *Histochem. J.* 4:331.

Bowes, J. H., and C. W. Cater. (1966) The reaction of glutaraldehyde with proteins and other biological materials. *J. Roy. Micros. Soc.* 85:193.

—— and C. W. Cater. (1968) The interaction of aldehydes with collagen. *Biochem. Biophys. Acta* 168:341.

—— and R. H. Kenten. (1949) The effect of deamination and esterification on the reactivity of collagen. *J. Biochem.* 44:142.

—— and A. H. Raistrick. (1964) The action of heat and moisture on leather. V. Chemical changes in collagen and tanned collagen. *J. Am. Leather Chem. Assoc.* 59:201.

——, C. W. Cater, and M. J. Ellis. (1965) Determination of formaldehyde and glutaraldehyde bound to collagen by carbon-14 assay. *J. Am. Leather Chem. Assoc.* 60:275.

Boyde, A., and E. Maconnachie. (1979) Volume changes during preparation of mouse embryonic tissue for scanning electron microscopy. *Scanning* 2:149.

—— and P. Vesely. (1972) Comparison of fixation and drying procedures for preparation of some cultured cell lines in the SEM. *IITRI/SEM*, p. 265.

——, E. Bailey, S. J. Jones, and A. Tamarin. (1977) Dimensional changes during specimen preparation for scanning electron microscopy. *IITRI/SEM* 1:507.

Breathnach, A. S., and M. Martin. (1976) A comparison of membrane fracture faces of fixed and unfixed glycerinated tissue. *J. Cell Sci.* 21:437.

Brunk, U., P. Bell, P. Colling, N. Forsby, and B. A. Frederickson. (1975) SEM of in vitro cultured cells, osmotic effects during fixation. *IITRI/SEM*, p. 379.

Buchanan, G. M., and D. A. Stewart. (1980) Application of Spurr, LX112, LX112/Araldite 502, Poly/Bed 812, and Effapoxy embedding media to in vitro agar-cultured bone marrow colonies. *Proc. 38th Ann. EMSA Meet.*, p. 646.

Buckley, I. K. (1973a) Studies in fixation for electron microscopy using cultured cells. *Lab. Invest.* 29:398.

—— (1973b) The lysosomes of cultured chick embryo cells. *Lab. Invest.* 29:411.

Bullivant, S., and J. Hotchin. (1960) Chromyl chloride, a new stain for electron microscopy. *Exp. Cell. Res.* 21:211.

Burke, C. N., and C. W. Geiselman. (1971) Exact anhydride epoxy percentages for electron microscope embedding. *J. Ultrastr. Res.* 36:119.

Burkl, W., and H. Schiechl. (1968) A study of osmium tetroxide fixation. *J. Histochem. Cytochem.* 16:157.

Bushman, R. J., and A. B. Taylor. (1968) Extraction of absorbed lipid (linoleic acid-1-^{14}C) from rat intestinal epithelium during processing for electron microscopy. *J. Cell Biol.* 38:252.

Busson-Mabillot, A. (1971) Influence de la fixation chimique sur les ultrastructures. *J. Microscopie* 12:317.

Butler, T. C., W. J. Waddell, and D. T. Poole. (1967) Intracellular pH based on the distribution of weak electrodes. *Fed. Proc.* 26:1327.

Callahan, W. P., and J. A. Horner. (1964) The use of vanadium as a stain for electron microscopy. *J. Cell Biol.* 20:350.

Carstensen, E. L., W. G. Aldridge, S. Z. Child, and P. Sullivan. (1971) Stability of cells fixed with glutaraldehyde and acrolein. *J. Cell Biol.* 50:529.

Casley-Smith, J. R. (1967) Some observations on the electron microscopy of lipids. *J. Roy. Micros. Soc.* 87:463.

Caulfield, J. B. (1957) Effects of varying the vehicle for osmium tetroxide in tissue fixation. *J. Biophys. Biochem. Cytol.* 3:827.

Chambers, R. W., M. C. Bowling, and P. M. Grimley. (1968) Glutaraldehyde fixation in routine histopathology. *Arch. Path.* 85:18.

Chang, S. C. (1971) Compounding of Luft's Epon embedding medium for use in electron microscopy with reference to anhydride: epoxide ratio adjustment. *Mikroskopie* 29:337.

Chapman, D., and D. J. Fluck. (1966) Physical properties of phospholipids. III. Electron microscope studies of some pure fully saturated 2,3-diacyl-dl-phosphatidyl-ethanolamines and phosphatidyl-cholines. *J. Cell Biol.* 30:1.

Charret, R., and E. Bauré-Fremiet. (1967) Technique de rassemblement de microorganismes: Préinclusion dans un caillot de fibrine. *J. Microscopie* 6:1063.

Chien, K. (1980) In situ embedding of cell monolayers on untreated glass surfaces for vertical and horizontal ultramicrotomy. *Proc. 38th Am. EMSA Meet.*, p. 644.

Chisalita, D., G. Chirita, and G. Alexa. (1971) Kinetics of dialdehyde combination with the reactive groups of collagen. *Ind. Usoara* 18:269.

Claude, A. (1961) Problems of fixation for electron microscopy. Results of fixation with osmium tetroxide in acid and alkaline media. *Pathol. Biol.* 9:933.

Cohen, A. L. (1979) Critical point-drying–principles and procedures. *SEM, Inc.* 2:303.

Collins, R. J., W. P. Griffith, F. L. Phillips, and A. C. Skapski. (1974) Staining and fixation of unsaturated membrane lipids by osmium tetroxide. *Biochem. Biophys. Acta* 354:152.

——, J. Jones, and W. P. Griffith. (1974) Reaction of osmium tetroxide with alkenes, glycols, and alkynes; oxo-osmium (VI) esters and their structures. *J. Chem. Soc.* (*A*), 1094.

Collins, V. P., U. Brunk, B. A. Fredericksson, and B. Westermark. (1980)

Transmission and scanning electron microscopy of whole glioma cells cultured in vitro. *SEM, Inc.* 2:223.
Conger, A., J. H. Garcia, A. S. Lossinsky, and F. C. Kauffman. (1978) The effect of aldehyde fixation on selected substrates for energy metabolism and amino acids in mouse brain. *J. Histochem. Cytochem.* 26:423.
Coulter, H. D. (1967) Rapid and improved methods for embedding biological tissue in Epon 812 and Araldite 502. *J. Ultrastr. Res.* 20:346.
Criegee, R. (1936) Osmiumsäure-ester als Zwischenprodukte bei Oxydationen. *Justus Liebigs Ann. Chem.* 522:75.
—— (1938) Organische Osmiumverbindungen. *Angew. Chem.* 51:519.
——, B. Marchand, and A. Wannowius. (1942) Zur Kenntnis der organischen Osmiumvergindungen. *Justus Liebigs Ann. Chem.* 550:99.
Czarnecki, C. M. (1971) The effect of fixation on the chemical extraction of glycogen from rat liver. *Histochem. J.* 3:163.
Darley, J. J., and H. Ezoe. (1976) Potential hazards of uranium and its compounds in electron microscopy: A brief review. *J. Micros.* 106:85.
Davey, D. F. (1973) The effect of fixative tonicity on the myosin filament lattice volume of frog muscle fixed following exposure to normal or hypertonic Ringer. *Histochem. J.* 5:87.
Dawson, R. M. C., D. C. Elliot, W. H. Elliot, and K. M. Jones. (1969) *Data for Biochemical Research*, 2nd ed. Clarendon Press, Oxford.
De Bruijn, W. C. (1973) Glycogen, its chemical and morphologic appearance in the electron microscope. I. A modified osmium tetroxide fixative which selectively contrasts glycogen. *J. Ultrastr. Res.* 42:29.
Dellman, H. D., and C. B. Pearson. (1977) Better epoxy resin embedding for electron microscopy at low relative humidity. *Stain Technol.* 52:5.
Demsey, A., D. Kawka, and C. W. Stackpole. (1978) Cell surface membrane organization revealed by freeze-drying. *J. Ultrastr. Res.* 62:13.
de Petris, S. (1965) Ultrastructure of the cell wall of *Escherichia coli. J. Ultrastr. Res.* 12:247.
Deter, R. L. (1973) Electron microscopic evaluation of subcellular fractions obtained by ultracentrifugation. In: *Principles and Techniques of Electron Microscopy* 3:201. Hayat, M. A. (ed.). Van Nostrand Reinhold, New York.
Deutsch, K., and H. Hillman. (1977) The effect of six fixatives on the areas of rabbit neurons and rat cerebral slices. *J. Micros.* 109:303.
Dreher, K. D., J. H. Schulman, O. R. Anderson, and O. A. Roels. (1967) The stability and structure of mixed lipid monolayers and bilayers. *J. Ultrastr. Res.* 19:586.
Ealker, J. F. (1964) *Formaldehyde*, 3rd ed. Chapman and Hall, London.
Elleder, M., and Z. Lojda. (1968a) Remarks on the detection of osmium derivatives in tissue sections. *Histochemie* 13:276.
—— and Z. Lojda (1968b) Remarks on the "OTAN" reaction. *Histochemie* 14:47.
Epling, G. P., and F. S. Sjostrand. (1962) In vitro and in vivo fixation of rabbit retina for electron microscopy. *Proc. 5th Int. Cong. EM*, P-5.
Eyring, E. J., and J. Ofengand. (1967) Reaction of formaldehyde with hetero-

cyclic imino nitrogen of purine and pyrimidine nucleosides. *Biochemistry* 6:2500.

Fahimi, H. D., and P. Drochmans. (1965a) Essais de standardization de la fixation au glutaraldehyde. I. Purification et determination de la concentration du glutaraldehyde. *J. Microscopie* 7:725.

—— and P. Drochmans. (1965b) Essais de standardization de la fixation au glutaraldehyde. II. Influence des concentrations en aldehyde et de l'osmolalite. *J. Microscopie* 4:737.

Falk, R. H. (1980) Preparation of plant tissues for SEM. *SEM, Inc.* 2:79.

Farquhar, M. G., and G. E. Palade (1965) Cell junctions in amphibian skin. *J. Cell Biol.* 26:263.

Farrant, J. L., and J. D. McLean. (1969) Albumins as embedding media for electron microscopy. *Proc. 27th Ann. EMSA Meet.*, p. 422.

Favard, P., N. Carasso, N. Bourguet, and S. Jard. (1966) Modifications provoqueés par la fixation et la déshydration sur la perméabilitie de la vessie de grenouille. *Proc. 6th Int. Cong. EM* 2:23.

Fawcett, D. W. (1966) *The Cell: Its Organelles and Inclusions.* W. B. Saunders Co., Philadelphia.

Feder, N. (1960) Some modification in conventional techniques of tissue preparation. *J. Histochem. Cyctochem.* 19:23A.

—— and T. P. O'Brien. (1968) Plant microtechnique: Some principles and new methods. *Am. J. Bot.* 55:123.

Finck, H. (1960) Epoxy resins in electron microscopy. *J. Biophys. Biochem. Cytol.* 7:27.

Fleischer, S., B. Fleischer, and W. Stoeckenius. (1967) Fine structure of lipid depleted mitochondria. *J. Cell Biol.* 32:193.

Flitney, E. W. (1966) The time course of fixation by formaldehyde, glutaraldehyde, acrolein, and other higher aldehydes. *J. Roy. Micros. Soc.* 85:353.

Forbes, M., and N. Sperelakis. (1971) Ultrastructure of lizard ventricular muscle. *J. Ultrastr. Res.* 34:439.

Franke, W. W., S. Krien, and R. M. Brown, Jr. (1969) Simultaneous glutaraldehyde-osmium tetroxide fixation with postosmication. *Histochemie* 19:162.

Furtado, J. S. (1970) The fibrin clot: A medium for supporting loose cells and delicate structures during processing for microscopy. *Stain Technol.* 45:19.

Gale, J. B. (1977) Differential effects of fixatives, buffers, and ionic species on the ultrastructure of heart mitochondria from resting and exhausted rats. *J. Electron Micros.* 26(2):185.

Geiselman, C. W., and C. N. Burke. (1973) Exact anhydride: epoxy percentages for Araldite and Araldite-Epon embedding, *J. Ultrastr. Res.* 43:220.

Gertz, S. D., M. L. Rennels, M. S. Forbes, and E. Nelson. (1975) Preparation of vascular endothelium for scanning electron microscopy: A comparison of the effects of perfusion and immersion fixation. *J. Micros.* 105:309.

Gibbons, I. R. (1958) A water-miscible embedding resin for electron microscopy. *Proc. 4th Int. Cong. EM (Berlin)* 2:55.

—— (1959) An embedding resin miscible with water for electron microscopy. *Nature (London)* 184:375.

Gil, K. H., and E. R. Weibel. (1968) The role of buffers in lung fixation with glutaraldehyde and OsO_4. *J. Ultrastr. Res.* 25:331.

—— and E. R. Weibel. (1969–70) Improvements in demonstration of lining layer of lung alveoli by electron microscopy. *Resp. Physiol.* 8:13.

Gillett, R., G. E. Jones, and T. Partridge. (1975) Distilled glutaraldehyde: Its use in an improved fixation regime for cell suspensions. *J. Micros.* 105:325.

Glauert, A. M. (1967) The fixation and embedding of biological samples. In: *Techniques for Electron Microscopy*, 2nd ed., p. 166. Kay, D. H. (ed.). Blackwell Scientific Publ., Oxford.

—— (1975) Fixation, dehydration, and embedding of biological specimens. In: *Practical Methods in Electron Microscopy*, vol. 3, part 1. Glauert, A. M. (ed.). American Elsevier, New York.

—— and R. H. Glauert. (1958) Araldite as an embedding medium for electron microscopy. *J. Biophys. Biochem. Cytol.* 4:409.

—— and M. J. Thornley. (1966) Glutaraldehyde fixation of Gram-negative bacteria. *J. Roy. Micros. Soc.* 85:449.

——, G. E. Rogers, and R. H. Glauert. (1956) A new embedding medium for electron microscopy. *Nature* 178:803.

Goldfarb, D., K. Miyai, and J. Hegenauer. (1977) A hydrostatic device for tissue dehydration. *Stain Technol.* 52:171.

Gomori, G. (1946) Buffers in the range of 6.5 to 9.6. *Proc. Soc. Exp. Biol. Med.* 62:33.

—— (1955) In: *Methods in Enzymology* 1:142. Colowick, S. P., and N. O. Caplan (eds.). Academic Press, New York.

Gonzalez-Aguilar, F. (1969) Extracellular space in the rat brain after fixation with 12 M formaldehyde. *J. Ultrastr. Res.* 29:76.

Good, N. E., G. D. Winget, W. Winter, T. Connolly, S. Izawa, and R. M. M. Singh. (1966) Hydrogen ion buffers for biological research. *Biochemistry* 5:467.

Gorycki, M. A. (1978) Mixing embedding media in plastic bags. *Stain Technol.* 53:116.

——, and V. Askanas. (1977) Improved technique for electron microscopy of cultured cells. *Stain Technol.* 52:249.

Griffith, W. P., A. C. Skapski, K. A. Woode, and M. J. Wright. (1978) Partial coordination in amine adducts of osmium tetroxide: X-ray molecular structure of quinuclidinetetraoxo-osmium (VII). *Inorg. Chim. Acta* 31:L413.

Grossenbacher, K. A., and R. Handley. (1969) Hardness of Maraglass as affected by solvent-monomer interaction. *Stain Technol.* 44:191.

Grund, S., J. Eichberg, and W. Engel. (1978) Biologische Strukturen im Photoelektronenmissionmikroskop. Semidünnschnite von Kükeneleler in den Einbettungsmittein Durcupan, Vestopal, Araldit, und Epon. *J. Utrastr. Res.* 64:191.

Gusnard, D., and R. H. Kirschner. (1977) Cell and organelle shrinkage during preparation for scanning electron microscopy: Effects of fixation, dehydration, and critical point drying. *J. Micros.* 110:51.

Habeeb, A. F. S. A., and R. Hiramoto. (1968) Reaction of proteins with gluatraldehyde. *Arch. Biochem. Biophys.* 126:16.

Hagstrom, L., and G. F. Bahr. (1960) Penetration rates of osmium tetroxide with different fixation vehicles. *Histochemie* 2:1.

Hake, T. (1965) Studies on the reactions of OsO_4 and $KMnO_4$ with amino acids, peptides, and proteins. *Lab. Invest.* 14:470.

Hampton, J. C. (1965) Effects of fixation on the morphology of Paneth cell granules. *Stain Technol.* 40:283.

Haselkorn, R., and P. Doty. (1961) The reaction of formaldehyde with polynucleotides. *J. Biol. Chem.* 236:2738.

Hayat, M. A. (1968) Triple fixation for electron microscopy. *Proc. 26th Ann. EMSA*, p. 90.

—— (1969) Uranyl acetate as a stain and a fixative for heart tissue. *Proc. 27th Ann. EMSA*, p. 412.

—— (1970) *Principles and Techniques of Electron Microscopy*, vol. I, Van Nostrand Reinhold, New York.

—— (1973) Specimen preparation. In: *Electron Microscopy of Enzymes* 1:1. Hayat, M. A. (ed.). Van Nostrand Reinhold, New York.

—— (1975) *Positive Staining for Electron Microscopy.* Van Nostrand Reinhold, New York.

—— and R. Giaquinta. (1970) Rapid fixation and embedding for electron microscopy. *Tissue & Cell* 2:191.

Heckman, C. A., and R. J. Barnett. (1973) GACH: A water-miscible, lipid-retaining polymer for electron microscopy. *J. Ultrastr. Res.* 42:156.

Hirsch, J. G., and M. E. Fedorko. (1968) Ultrastructure of human leukocytes after simultaneous fixation with glutaraldehyde and osmium tetroxide and "postfixation" in uranyl acetate. *J. Cell Biol.* 38:615.

Hobbs, M. J. (1969) Fixation of microscopic fresh-water green algae by 31% OsO_4 in CCl_4 in an unbuffered, two-phase fixative system. *Stain Technol.* 44:217.

Hopwood, D. (1967a) Some aspects of fixation with glutaraldehyde: A biochemical and histochemical comparison of the effects of formaldehyde and glutaraldehyde fixation on various enzymes and glycogen, with a note on penetration of glutaraldehyde into liver. *J. Anat.* 101:83.

—— (1967b) The behavior of various glutaraldehydes on Sephadex G-10 and some implications for fixation. *Histochemie* 11:289.

—— (1969a) Fixatives and fixation: A review. *Histochem. J.* 1:323.

—— (1969b) Fixation of proteins by osmium tetroxide, potassium dichromate, and potassium permanganate. *Histochemie* 18:250.

—— (1970a) Use of isoelectric focusing to determine the isoelectric point of bovine serum albumin after treatment with various common fixatives. *Histochem. J.* 3:201.

—— (1970b) The reactions between formaldehyde, glutaraldehyde, osmium tetroxide and their fixation effects on bovine serum albumin and on tissue blocks. *Histochemie* 24:50.

—— (1972) Theoretical and practical aspects of glutaraldehyde fixation. *Histochem. J.* 4:267.

—— (1975) The reactions of glutaraldehyde with nucleic acids. *Histochem. J.* 7:267.

——, C. R. Allen, and M. McCabe. (1970) The reactions between glutaraldehyde and various proteins. An investigation of their kinetics. *Histochem. J.* 2:137.

Howard, K. S., and M. T. Postek. (1979) Dehydration of scanning electron microscopy specimens—a bibliography. *SEM, Inc.* 2:892.

Humphreys, W. J. (1977) Health and safety hazards in the SEM laboratory. *IITRI/SEM* 1:537.

Huxley, H. E., and G. Zubay. (1961) Preferential staining of nucleic acid-containing structures for electron microscopy. *J. Biophys. Biochem. Cytol.* 11:273.

Idelman, S. (1964) Modification de la technique de Luft en vue de la conservation des lipides en microscopie electronique. *J. Microscopie* 3:715.

—— (1965) Conservation des lipides par les techniques utilisées en microscopie electronique. *Histochemie* 5:18.

Igbal, S. J., and B. S. Weakley. (1974) The effects of different preparative procedures on the ultrastructure of the hamster ovary. *Histochem. J.* 38:95.

Jansen, E. F., Y. Tominatsu, A. and C. Olsen. (1971) Cross-linking of α-chymotrypsin and other proteins by reaction with glutaraldehyde. *Arch Biochem. Biophys.* 144:394.

Johnson, J. E., Jr. (1978) Transmission and scanning electron microscope preparations of the same cell culture. *Stain Technol.* 53:273.

Johnston, P. V. and B. I. Roots (1967) Fixation of the central nervous system with aldehydes and its effect on the extracellular space as seen by electron microscopy. *J. Cell Sci.* 2:377.

Johnston, W. H., H. Latta, and L. Osvaldo. (1973) Variations in glomerular ultrastructure in rat kidneys fixed by perfusion. *J. Ultrastr. Res.* 45:149.

Jones, D. (1972) Reactions of aldehydes with unsaturated acids during histological fixation. *Histochem. J.* 4:421.

—— and G. A. Gresham. (1966) Reaction of formaldehyde with unsaturated fatty acids during histological fixation. *Nature* 210:1386.

Jones, G. J. (1974) Polymerization of glutaraldehyde at fixative pH. *J. Histochem. Cytochem.* 22:911.

Jost, P., U. J. Brooks, and O. H. Griffith. (1973) Fluidity of phospholipid bilayers and membranes after exposure to osmium tetroxide and glutaraldehyde. *J. Mol. Biol.* 76:313.

Jurand, A., and M. Ireland. (1965) A slow rotary shaker for embedding in viscous media. *Stain Technol.* 40:233.

Kahn, L. E., S. P. Frommes, and P. A. Cancilla. (1977) Comparison of ethanol and chemical dehydration methods for the study of cells in culture by scanning and transmission electron microscopy. *IITRI/SEM* 1:501.

Kalimo, H. (1976) The role of the blood-brain barrier in perfusion fixation of the brain for electron microscopy. *Histochem. J.* 8:1.

Karlsson, U., and R. L. Schultz. (1965) Fixation of the central nervous system for electron microscopy by aldehyde perfusion. I. Preservation with aldehyde perfusates versus direct perfusion with osmium tetroxide with special reference to membrane and the extracellular space. *J. Ultrastr. Res.* 12:160.

Karnovsky, M. J. (1965) A formaldehyde–glutaraldehyde fixative of high osmolarity for use in electron microscopy. *J. Cell Biol.* 27:137A.

Kellenberger, E. A., J. Ryter, and J. Sechaud. (1958) Electron microscopy study of DNA-containing plasmas. II. Vegetative and mature phage DNA as compared with normal bacterial nucleoids in different physiological states. *J. Biophys. Biochem. Cytol.* 4:671.

Kiernan, J. A. (1978) Recovery of osmium tetroxide from used fixative solutions. *J. Micros.* 113:77.

Kuran, H., and M. J. Olzewska. (1977) Effects of some buffers on the ultrastructure, dry mass content, and radioactivity of nuclei of *Haemanthus Katharinea. Micros Acta* 79:69.

Kushida, H. (1959) On an epoxy resin embedding method for ultrathin sectioning. *J. Electron Micros.* 8:72.

—— (1960) A new polyester embedding method for ultrathin sectioning. *J. Electron Micros.* 9:157.

—— (1963) A modification of the water-miscible epoxy resin "Durcupan" embedding method for ultrathin sectioning. *J. Electron Micros.* 12:71.

—— (1964a) Improved method for embedding with Durcupan. *J. Electron Micros.* 13:139.

—— (1964b) Glycol methacrylate as a dehydrating agent for embedding with polyester and epoxy resins. *J. Electron Micros.* 13:200.

—— (1965) Durcupan as a dehydrating agent for embedding with polyester, styrene, and methacrylate resins. *J. Electron Micros.* 14:52.

—— (1966) Further improved methods for embedding with Durcupan. *J. Electron Micros.* 15:95.

—— (1969) A new rotary shaker for fixation, dehydration, and embedding. *J. Electron Micros.* 18:137.

Landis, W. J., M. C. Paine, and M. J. Glimcher. (1980) Use of acrolein vapors for the preparation of bone tissue for electron microscopy. *J. Ultrastr. Res.* 70:171.

Langenberg, W. G. (1979) Chilling of tissues before glutaraldehyde fixation preserves fragile inclusions of several plant viruses. *J. Ultrastr. Res.* 66:120.

Larsson, L. (1975) Effects of different fixatives on the ultrastructure of the developing proximal tubules in the rat kidney. *J. Ultrastr. Res.* 51:140.

Lawton, J., and P. J. Harris. (1978) Fixation of senescing plant tissues: Sclerenchymatous fibre cells from the flowering stem of a grass. *J. Micros.* 112:307.

Leduc, E. H., and W. Bernhard. (1967) Recent modifications of the glycol methacrylate embedding procedure. *J. Ultrastr. Res.* 19:196.

—— and S. J. Holt. (1965) Hydroxypropyl methacrylate, a new water-miscible embedding medium for electron microscopy. *J. Cell Biol.* 26:137.

——, V. Marinozzi, and W. Bernhard. (1963) The use of water soluble glycol methacrylate in ultrastructural cytochemistry. *J. Roy. Micros. Soc.* 81:119.

Levy, W. A., I. Herzog, K. Suzuki, R. Katzman, and L. Schienberg. (1965)

Method for combined ultrastructural and biochemical analysis of normal tissue. *J. Cell Biol.* 27:119.

Lin, C. H., R. H. Falk, and C. R. Stocking. (1977) Rapid chemical dehydration of plant material for light and electron microscopy with 2,2-dimethoxypropane and 2,2-diethoxypropane. *Am. J. Bot.* 64:602.

Litman, R. B., and R. Barrnett. (1972) The mechanism of the fixation of tissue components by osmium tetroxide via hydrogen bonding. *J. Ultrastr. Res.* 38:63.

Lojda, A. (1965) Fixation in histochemistry. *Folia Morph.* 13:65.

Lombardi, L., G. Prenna, L. Okdicsanyi, and A. Gautier. (1971) Electron staining with uranyl acetate. Possible role of free amino groups. *J. Histochem. Cytochem.* 19:161.

Luft, J. H. (1959) The use of acrolein as fixative for light and electron microscopy. *Anat. Rec.* 133:305.

—— (1961) Improvements in epoxy resin embedding methods. *J. Biophys. Biochem. Cytol.* 9:409.

—— (1966) Ruthenium red staining of the striated muscle cell membrane and the myotendinal junction. *Proc. 6th Int. Cong. EM (Tokyo)* 2:65.

Machado, A. B. M. (1967) Straight OsO_4 versus glutaraldehyde–OsO_4 in sequence as fixatives for the granular vesicles in sympathetic axons of the rat pineal body. *Stain Technol.* 42:293.

Malhotra, S. K. (1962) Experiments on fixation for electron microscopy. I. Unbuffered osmium tetroxide. *Quart. J. Micros. Sci.* 103:5.

Marinozzi, V. (1961) Silver impregnation of ultrathin sections for electron microscopy. *J. Biophys. Biochem. Cytol.* 9:121.

—— (1963) The role of fixation in electron staining. *J. Roy. Micros. Soc.* 81:141.

Maser, M. D., T. E. Powell, and C. W. Philpott. (1967) Relationships among pH, osmolarity, and concentration of fixative solution. *Stain Technol.* 42:175.

Maser, M. M., and J. J. Trimble. (1977) Rapid chemical dehydration of biological samples for scanning electron microscopy using 2,2-dimethoxypropane. *J. Histochem. Cytochem.* 25:247.

Mathieu, O., H. Classen, and E. R. Weibel. (1978) Differential effect of glutaraldehyde and buffer osmolarity on cell dimensions: A study of lung tissue. *J. Ultrastr. Res.* 63:20.

Maunsbach, A. B. (1966) The influence of different fixatives and fixation methods on the ultrastructure of rat kidney proximal tubule cells. II. Effects of varying osmolarity, ionic strength, buffer systems, and fixative concentration of glutaraldehyde solutions. *J. Ultrastr. Res.* 15:283.

——, S. C. Madden, and H. Latta. (1962) Variations in fine structure of renal epithelium under different conditions of fixation. *J. Ultrastr. Res.* 6:511.

McLean, J. D. (1960) Fixation of plant tissue. *Proc. 4th Int. Cong. EM (Berlin)* 2:27.

Mersey, B., and M. E. McCully. (1978) Monitoring of the course of fixation of plant cells. *J. Micros.* 114:49.

Millonig, G. (1961) Advantages of a phosphate buffer for osmium tetroxide solutions in fixation. *J. Appl. Phys.* 32:1637.

—— (1964) Model experiments on fixation and dehydration. *Proc. 6th Int. Cong. EM (Tokyo)* 2:21.

—— and V. Marinozzi. (1968) Fixation and embedding in electron microscopy. In: *Advances in Optical and Electron Microscopy* 2:251. Barer, R., and V. E. Cosslett (eds.). Academic Press, New York.

—— (1964) Plastic embedding mixtures for use in electron microscopy. *Stain Technol.* 39:111.

Moretz, R. C., C. K. Akers, and D. F. Parsons. (1969a) Use of small angle x-ray diffraction to investigate disordering of membranes during preparation for electron microscopy. I. Osmium tetroxide and potassium permanganate. *Biochem. Biophys. Acta* 193:1.

——, C. K. Akers, and D. F. Parsons. (1969b) Use of small angle x-ray diffraction to investigate disordering of membranes during preparation for electron microscopy. II. Aldehydes. *Biochem. Biophys. Acta* 193:12.

Morgan, T. E., and G. L. Huber. (1967) Loss of lipid during fixation for electron microscopy. *J. Cell Biol.* 32:757.

Morre, D. J., and H. H. Mollenhauer. (1969) Studies on the mechanism of glutaraldehyde stabilization of cytoplasmic membranes. *Proc. Ind. Acad. Sci.* 78:167.

Moss, G. I. (1966) Glutaraldehyde as a fixative for plant tissues. *Protoplasma* 62:194.

Muller, L. L., and T. J. Jacks. (1975) Rapid chemical dehydration of samples for electron microscope examination. *J. Histochem. Cytochem.* 23:107.

Mumaw, V. R., and B. C. Munger. (1971) Uranyl acetate as a fixative from pH 2.0 to 8.0 *Proc. 29th Ann. EMSA Meet.*, p. 490.

Nelson, B. K., and B. A. Flaxman. (1972) In situ embedding and vertical sectioning for electron microscopy of tissue cultures grown in plastic Petri dishes. *Stain Technol.* 47:261.

Newman, S. B., E. Borysko, and M. Swerdlow. (1949) New sectioning techniques for light and electron microscopy. *Science* 110:66.

Nielson, A. J., and W. P. Griffith. (1978) Tissue fixation and staining with osmium tetroxide: The role of phenolic compounds. *J. Histochem. Cytochem.* 26:138.

—— and W. P. Griffith. (1979) Tissue fixation by osmium tetroxide: A possible role for proteins. *J. Histochem. Cytochem.* 27:997.

Nopanitaya, W., R. K. Charlton, R. C. Turchis, and J. W. Grisham. (1977) Ultrastructure of cells cultured on polycarbonate membranes. *Stain Technol.* 52:143.

Norton, T. N., M. Gelfand, and M. Brotz. (1962) Studies in the histochemistry of plasmalogens: I. The effect of formalin and acrolein fixation on the plasmalogens of adrenal and brain. *J. Histochem. Cytochem.* 10:375.

Nowell, J. A., and J. B. Pawley (1980) Preparation of experimental animal tissue for SEM. *SEM, Inc.* 2:1.

O'Brien, T. P., J. Kuo, M. E. McCully, and S. Y. Zee. (1973) Coagulant and noncoagulant fixation of plant cells. *Aust. J. Biol. Sci.* 26:1231.
Ockelford, C. D. (1975) Redundancy of washing in the preparation of biological specimens for transmission electron microscopy. *J. Micros.* 105:193.
Page, S. G., and H. E. Huxley. (1963) Filament lengths in striated muscle. *J. Cell Biol.* 19:369.
Palade, G. E. (1952) A study of fixation for electron microscopy. *J. Exp. Med.* 95:285.
Palay, S. L., S. M. McGee-Russell, S. Gordon, and M. A. Grillo. (1962) Fixation of neutral tissue for electron microscopy by perfusion with solutions of osmium tetroxide. *J. Cell Biol.* 12:385.
Parsons, D. F., E. B. Bole, D. J. Hall, and W. D. E. Thomas. (1974) A comparative survey of techniques for preparing plant surface for the scanning electron microscope. *J. Micros.* 101:59.
Pease, D. C. (1973) Glycol methacrylate copolymerized with glutaraldehyde and urea as an embedment retaining lipids. *J. Ultrastr. Res.* 45:124.
—— (1980) Ultrathin sectioning of fixed but unembedded tissues. *Proc. 38th Ann. EMSA Meet.*, p. 650.
—— and R. G. Peterson. (1972) Polymerizable glutaraldehyde-urea mixtures as polar, water containing embedding media. *J. Ultrastr. Res.* 41:133.
Pentilla, A., H. Kalimo, and B. F. Trump. (1974) Influence of glutaraldehyde and/or osmium tetroxide on cell volume, ion content, mechanical stability, and membrane permeability of Ehrlich ascites tumor cells. *J. Cell Biol.* 63:197.
——, E. M. McDowell, and B. F. Trump. (1975) Effects of fixation and post-fixation treatments on volume of injured cells. *J. Histochem. Cytochem.* 23:251.
Peracchia, C., and B. S. Mittler. (1972) New glutaraldehyde fixation procedures. *J. Ultrastr. Res.* 39:57.
Perre, J., and J. Foncin. (1977) An embedding method for cell cultures for light and electron microscopy. *Stain Technol.* 52:240.
Peterson, R. G., and D. C. Pease. (1970a) Polymerized glutaraldehyde-urea mixtures as water-soluble embedding media. *Proc. 28th Ann. EMSA Meet.*, p. 334.
—— and D. C. Pease. (1970b) Features of the fine structure of myelin embedded in water-containing aldehyde resins. *Proc. 7th Int. Cong. EM (Grenoble)* 1:409.
—— and D. C. Pease. (1972) Myelin embedded in polymerized glutaraldehyde-urea. *J. Ultrastr. Res.* 41:115.
Pexieder, T. (1976) The role of buffer osmolarity in fixation for SEM and TEM. *Experientia* 32:806.
Pilstrom, L., and U. Nordlund. (1975) The effect of temperature and concentration of the fixative on morphometry of rat liver mitochondria and rough endoplasmic reticulum. *J. Ultrastr. Res.* 50:33.
Pladellorens, M., and J. A. Subirana. (1975) Preservation of membrane ultrastructure with aldehyde or imidate fixatives. *J. Ultrastr. Res.* 52:243.
Porter, K. R., and F. Kallman. (1953) The properties and effects of osmium

tetroxide as a tissue fixative with special reference to its use for electron microscopy. *Exp. Cell Res.* 4:127.

Quiocho, F. A., W. H. Bishop, and F. M. Richards. (1967) Effects of changes in some solvent parameters on carboxypeptidase A in solution and in cross-linked crystals. *Proc. Nat. Acad. Sci.* 57:525.

Reedy, M. K. (1965) Section staining for electron microscopy. *J. Cell Biol.* 26:309.

Reimer, L. (1959) Quantitative untersuchung zur Massenbnahme von Einbetturgsmittein (Methacrylat, Vestopal, und Araldit) unter Elektronenbeschuss. *Z. Naturforsch.* 146:566.

Rhodin, J. (1954) Correlation of ultrastructural organization and function in normal and experimentally changed proximal convoluted tubule cells of the mouse kidney. Thesis, Karolinska Institutet, Stockholm, Aktiebolaget Godvil.

Richards, F. M., and J. R. Knowles. (1968) Glutaraldehyde as a protein cross-linking reagent. *J. Mol. Biol.* 37:321.

Riemersma, J. C. (1968) Osmium tetroxide fixation of lipids for electron microscopy: A possible reaction mechanism. *Biochim. Biophys. Acta* 152:718.

—— (1970) Chemical effects of fixation on biological specimens. In: *Some Biological Techniques in Electron Microscopy*, p. 69. Parsons, D. F., (ed.). Academic Press, New York.

—— and H. L. Bouijn. (1962) The reaction of osmium tetroxide with lecithin: Application of staining procedures. *J. Histochem. Cytochem.* 10:89.

Robertson, E. A., and R. L. Schultz. (1970) The impurities in commercial glutaraldehyde and their effect on the fixation of brain. *J. Ultrastr. Res.* 30:275.

Robertson, J. G., P. Lyttleton, K. I. Williamson, and R. D. Batt. (1975) The effect of fixation procedures on the electron density of polysaccharide granules in *Nocardia corallina. J. Ultrastr. Res.* 52:321.

Rodriguez-Garcia, M. I., and J. C. Stockert. (1979) Localization of the pyro-antimonate-osmium reaction product in plant cell nucleoli. *J. Ultrastr. Res.* 67:65.

Rosenberg, M., P. Bartl, and J. Lesko. (1960) Water-soluble methacrylate as an embedding medium for the preparation of ultrathin sections. *J. Ultrastr. Res.* 4:298.

Rosene, D. L., and M. M. Mesulam. (1978) Fixation variables in horseradish peroxidase neurohistochemistry. I. The effects of fixative and perfusion procedures upon enzyme activity. *J. Histochem. Cytochem.* 26:28.

Ryter, A., and E. Kellenberger. (1958a) Etude au microscope électronique de plasma contenant de l'acid désoxyribonucleique. I. Les nucléotides des bactéries en croissance active. *Z. Naturforsch.* 136:597.

—— and E. Kellenberger. (1958b) L'inclusion au polyester pour l'ultramicrotomie. *J. Ultrastr. Res.* 2:200.

Sabatini, D. D., K G. Bensch, and R. J. Barrnett. (1962) New fixatives for cytological and cytochemical studies. *Proc. 5th Int. Cong. EM (Philadelphia)* 2:1.

——, K. Bensch, and R. J. Barrnett. (1963) New means of fixation for electron microscopy and histochemistry. *Anat. Rec.* 142:274.

——, F. Miller, and R. J. Barrnett. (1964) Aldehyde fixation for morphological and enzyme histochemical studies with the electron microscope. *J. Histochem. Cytochem.* 12:57.

Saito, T., and H. Keino. (1976) Acrolein as a fixative for enzyme cytochemistry. *J. Histochem. Cytochem.* 24:1258.

Salema, R., and I. Brandao. (1973) The use of PIPES buffer in the fixation of plant cells for electron microscopy. *J. Submicro. Cytol.* 9:79.

Sandborn, E. (1966) Electron microscopy of the neuron membrane system and filaments. *Can. J. Physiol. Pharmacol.* 44:329.

Saubermann, A. J. (1980) Application of cryosectioning to x-ray microanalysis of biological tissue. *SEM, Inc.* 2:421.

Sawicki, W., and J. Lipitz. (1971) Albumin embedding and individual mounting of one or many mammalian ova on slides for fluid processing. *Stain Technol.* 46:261.

Sax, N. I. (1975) *Dangerous Properties of Industrial Materials*, 4th ed. Van Nostrand Reinhold, New York.

Schidlovsky, G. (1965) Contrast in multilayer system after various fixations. *Lab. Invest.* 14:1213.

Schiechl, H. (1971) Der chemismus der OsO_4-fixierung und sein eninfluss auf die zellstruktur. *Acta Histochem.* (*Suppl.*) 10:165.

Schiff, R., and J. F. Gennaro. (1979a) The influence of the buffer on maintenance of tissue lipids in specimens for scanning electron microscopy. *SEM, Inc.* 3:449.

—— and J. F. Gennaro, Jr. (1979b) The role of the buffer in the fixation of biological specimens for transmission and scanning electron microscopy. *Scanning* 2:135.

——, L. Grillone, D. Rutherford, and J. F. Gennaro. (1976) The influence of the buffer on maintenance of tissue lipid in specimens for SEM. *In Vitro* 12(4):305.

Schlatter, C., and I. Schlatter-Lanz. (1971) A simple method for the regeneration of used osmium tetroxide solutions. *J. Micros.* 94:85.

Schmalbruch, H. (1980) Delayed fixation alters the pattern of intramembrane particles in mammalian muscle fibrils. *J. Ultrastr. Res.* 70:15.

Schultz, R. L., and N. M. Case. (1968) Microtubule loss with acrolein and bicarbonate-containing fixatives. *J. Cell Biol.* 38:633.

——, and N. M. Case (1970) A modified aldehyde perfusion technique for preventing certain artifacts in electron microscopy of the central nervous system. *J. Micros.* 92:69.

—— and U. Karlsson. (1965) Fixation of the central nervous system for electron microscopy by aldehyde perfusion. II. Effect of osmolarity, pH of perfusate, and fixative concentration. *J. Ultrastr. Res.* 12:187.

Séchaud, J., and E. Kellenberger. (1972) Electron microscopy of DNA-containing plasms. Glutaraldehyde–uranyl acetate fixation of virus infected bacteria for thin sectioning. *J. Ultrastr. Res.* 39:598.

Shands, J. W. Jr. (1968) Embedding free-floating cells and microscopic particles: Serum albumin coagulum-epoxy resin. *Stain Technol.* 43:15.

Shay, J. W., and C. Walker (1980) Introduction to cells in culture as studied by SEM. *SEM, Inc.* 2:171.

Shinagawa, Y., Y. Shinagawa, and S. Uchida. (1980) Water containing Epon 812/815 embedding method for electron microscopy. *Proc. 38th Ann. EMSA Meet.*, p. 642.

Silva, M. T. (1973) Uranium salts. In: *Encyclopedia of Microscopy and Microtechnique*, p. 585. Gray, P. (ed.). Van Nostrand Reinhold, New York.

——, F. C. Guerra, and M. M. Magalhaes. (1968) The fixative action of uranyl acetate in electron microscopy. *Experientia* 24:1074.

——, J. M. S. Mota, J. V. C. Melo, and F. C. Guerra. (1971) Uranyl salts as fixatives for electron microscopy. Study of the membrane ultrastructure and phospholipid loss in bacilli. *Biochim. Biophys. Acta* 233:513.

Simson, J. A. V., R. M. Dom, P. L. Sannes, and S. S. Spicer. (1978) Morphology and cytochemistry of acinar secretory granules in normal and isoproterenol treated rat submandibular glands. *J. Micros.* 113:185.

Sjostrand, F. S. (1956) Electron microscopy of cells and tissues. In: *Physical Techniques in Biological Research* 3:241. Oster, G., and A. W. Pollister (eds.). Academic Press, New York.

—— and H. Halma. (1978) A centrifuge head for back-and-forth centrifugation to improve specimen infiltration by viscous plastics. *J. Ultrastr. Res.* 64:261.

Skaer, R. J., and S. Whytock. (1977) The fixation of nuclei in glutaraldehyde. *J. Cell Sci.* 27:13.

Spaur, R. C., and G. C. Moriarty. (1972) Improvements of glycol methacrylate. I. Its use as an embedding medium for electron microscopical studies. *J. Histochem. Cytochem.* 25:163.

Spurr, A. R. (1969) A low-viscosity epoxy resin embedding medium for electron microscopy. *J. Ultrastr. Res.* 26:31.

Staubli, W. (1960) Nouvelle matiére d'inclusion hydrosoluble pour la cytologie électronique. *C. R. Seanc. Soc. Biol.* 250:1137.

Stein, O., and Y. Stein. (1971) Lipid synthesis, intracellular transport, storage and secretion. I. Electron microscope autoradiographic study of liver after injection of tritiated palmitate or glycerol in fasted and ethanol-treated rats. *J. Cell Biol.* 33:319.

Steinbrecht, R. A., and K. D. Ernst. (1967) Continuous penetration of delicate tissue specimens with embedding resin. *Sci. Tools* 14:24.

Stoeckinius, W., and S. C. Mahr. (1965) Studies on the reaction of osmium tetroxide with lipids and related compounds. *Lab. Invest.* 14:458.

Strauss, E. W., and A. A. Arabian. (1969) Fixation of long-chain fatty acid in segments of jejunum from golden hamster. *J. Cell Biol.* 43:140a.

Sturgess, J. M., M. M. Mitranic, and M. A. Moscarello. (1978) Extraction of glycoproteins during tissue preparation for electron microscopy. *J. Micros.* 114:101.

Tahmisian, T. N. (1964) Use of the freezing point to adjust the tonicity of fixing solutions. *J. Ultrastr. Res.* 10:182.

Terzakis, J. A. (1968) Uranyl acetate, a stain and a fixative. *J. Ultrastr. Res.* 22:168.

Thornell, L. E., M. Sjöström, U. Karlsson, and E. Cedergren. (1977) Variable opacity of glycogen in routine electron micrography. *J. Histochem. Cytochem.* 25:1069.

Thornwaite, J. T., R. A. Thomas, S. B. Leif, T. A. Yopp, B. F. Cameron, and R. C. Leif. (1978) The use of electronic cell volume analysis with the AMAC II to determine the optimum glutaraldehyde fixative concentration for nucleated mammalian cells. *Sem, Inc.* 2:1123.

Thorpe, J. R., and D. M. R. Harvey. (1979) Optimization and investigation of the use of 2,2-dimethyoxypropane as the dehydration agent for plants in transmission electron microscopy. *J. Ultrastr. Res.* 68:186.

Thurston, E. L. (1978) Health and safety hazards in the SEM laboratory: Update 1978. *SEM, Inc.* 2:849.

Thurston, R. Y., R. A. Hess, K. H. Kilburn, and W. M. McKenzie. (1976) Ultrastructure of lung fixed in inflation using a new osmium fluorocarbon technique. *J. Ultrastr. Res.* 56:39.

Ting-Beall, H. P. (1980) Interactions of uranium ions with lipid bilayer membranes. *J. Micros.* 118:221.

Tobin, T. P. (1980) The osmotic effect of glutaraldehyde fixative components. *Proc. 38th Ann. EMSA Meet.*, p. 638.

Tomimatsu, Y., E. F. Jansen, W. Gaffield, and A. C. Olsen. (1971) Physical chemical observation on the α-chymotrypsin–glutaraldehyde system during formation of an insoluble derivation. *J. Colloid Interface Sci.* 36:51.

Tooze, J. (1964) Measurements of some cellular changes during the fixation of amphibian erythrocytes with osmium tetroxide solutions. *J. Cell Biol.* 22:551.

Tormey, J. McD. (1965) Artifactural localization of ferritin in the ciliary epithelium in vitro. *J. Cell Biol.* 25:1.

Trnavska, Z., S. Sitaj, M. Grmela, and J. Malinsky. (1966) Certain intermediary metabolites and the formation of fibrils from collagen solutions. *Biochim. Biophys. Acta* 126:373.

Trump, B. F., and R. E. Bulger. (1966) New ultrastructural characteristics of cells fixed in a glutaraldehyde–osmium tetroxide mixture. *Lab. Invest.* 15:368.

—— and J. L. E. Ericsson. (1965) The effect of the fixative solution on the ultrastructure of cells and tissues. A comparative analysis with particular attention to the proximal convoluted tubule of the rat kidney. *Lab. Invest.* 14:1245.

van der Wal, U. P., and M. R. Dohmen. (1978) A method for the orientation of small and delicate objects in embedding media for light and electron microscopy. *Stain Technol.* 53:56.

van Deurs, B., and J. H. Luft. (1979) Effects of glutaraldehyde fixation on the structure of tight junctions. A quantitative freeze fracture analysis. *J. Ultrastr. Res.* 68:160.

van Duijn, P. (1961) Acrolein–Schiff, a new staining method for proteins. *J. Histochem. Cytochem.* 9:234.

Van Harreveld, A., and F. I. Khattab. (1968) Perfusion fixation with glutaraldehyde and post-fixation with osmium tetroxide for electron microscopy. *J. Cell Sci.* 3:579.

Verwey, E. J., and J. T. G. Overbeek. (1948) *Theory of the Stability of Lyophobic Colloids.* Elsevier, Amsterdam.

Ward, B. J., and J. A. Gloster. (1976) Lipid losses during processing of cardiac muscle for electron microscopy. *J. Micros.* 108:41.

Watson, M. C., and W. C. Aldridge. (1961) Methods for the use of indium as an electron stain. *J. Biophys. Biochim. Cytol.* 11:257.

Weakly, B. S. (1974) A comparison of three different electron microscopical grade glutaraldehydes used to fix ovarian tissues. *J. Micros.* 101:127.

—— (1977) How dangerous is sodium cacodylate? *J. Micros.* 109:249.

West, J., and J. L. Mangan. (1970) Effects of glutaraldehyde on the protein loss and photochemical properties of kale chloroplasts: Preliminary studies on food conversion. *Nature (London)* 228:466.

White, D. L., S. B. Andrews, J. W. Faller, and R. J. Barrnett. (1976) The chemical nature of osmium tetroxide fixation and staining of membranes by x-ray photoelectron spectroscopy. *Biochim. Biophys. Acta* 436:577.

Willison, J. H. M., and R. Rajaraman. (1977) "Large" and "small" nuclear pore complexes: The influence of glutaraldehyde. *J. Micros.* 109:183.

Winborn, W. B. and L. L. Seelig (1970) Paraformaldehyde and s-collidine-a fixative for processing large blocks of tissue for electron microscopy. *Tex. Rep. Biol. Med.* 28:347.

Winters, C., and M. Slade. (1971) Embedding free-floating cells in stained agar for rapid screening and subsequent ultramicrotomy. *Stain Technol.* 46:161.

Wolfe, S. L., M. Beer, and C. R. Zobel. (1962) The selective staining of nucleic acids in a model system and in tissue. *Proc. 5th Int. Cong. EM (Philadelphia)* 2:0–6.

Wolman, M. (1955) Problems of fixation in cytology, histology, and histochemistry. *Int. Rev. Cytol.* 4:79.

—— and J. Greco. (1952) The effect of formaldehyde on tissue lipids and on histochemical reactions for carbonyl groups. *Stain Technol.* 29:317.

Wood, R. L., and J. H. Luft. (1963) The influence of the buffer system on fixation with osmium tetroxide. *J. Cell Biol.* 19:83A.

—— and J. H. Luft. (1965) The influence of buffer systems on fixation with osmium tetroxide. *J. Ultrastr. Res.* 12:22.

Wriggleworth, J. M. and L. Packer (1969) pH-dependent conformational change in submitochondrial particles. *Arch. Biochem. Biophys.* 133:194.

Zalokar, M., and I. Erk. (1977) Phase partition fixation and staining of *Drosophila* eggs. *Stain Technol.* 52:89.

Zeikus, J. A., and H. C. Aldrich. (1975) Use of hot formaldehyde fixation in processing plant-parasitic nematodes for electron microscopy. *Stain Technol.* 50:219.

Zobel, C. R., and M. Beer. (1961) Electron stains I. Chemical studies on the interaction of DNA with uranyl salts. *J. Biophys. Biochem. Cytol.* 10:335.

8. Thin Sectioning

INTRODUCTION

Ultramicrotomy refers to the production of sections that are thin enough to permit electron imaging in the TEM. Because the electron contrast of the section has been enhanced during osmium tetroxide and uranyl acetate fixation, electron-opaque (stained) areas will absorb or scatter the primary beam, whereas electron-transparent regions have minimal or no interaction with the beam. If specimen thickness exceeds ~800 Å, the primary beam will be absorbed (i.e., no signal will be transmitted). Consequently, embedded tissues must be thin-sectioned to distinguish electron-opaque from electron-transparent areas.

Table 8-1 is a flow chart showing the basic steps for sectioning. Following polymerization, the tissue blocks are removed from the embedding mold. When conventional BEEM capsules are used, the excess embedding medium immediately surrounding the tissue is removed by shaving with a razor; this is referred to as back-cutting. With flat embedding molds, the degree of back-cutting depends upon the sample orientation. Ultramicrotomes permit simultaneous specimen trimming and thick sectioning for light microscopy, which results in a much smoother block face, an asset for thin sectioning. Although the preparation of thick sections is discussed in Chapter 10, the reader should be aware that final specimen trimming is always done on the microtome.

The same specimen is then thin-sectioned for TEM using either a diamond knife (mounted in its own holder) or a glass knife. The stages of glass knife preparation include controlled knife breakage

Table 8-1. Flow Chart of Sectioning.

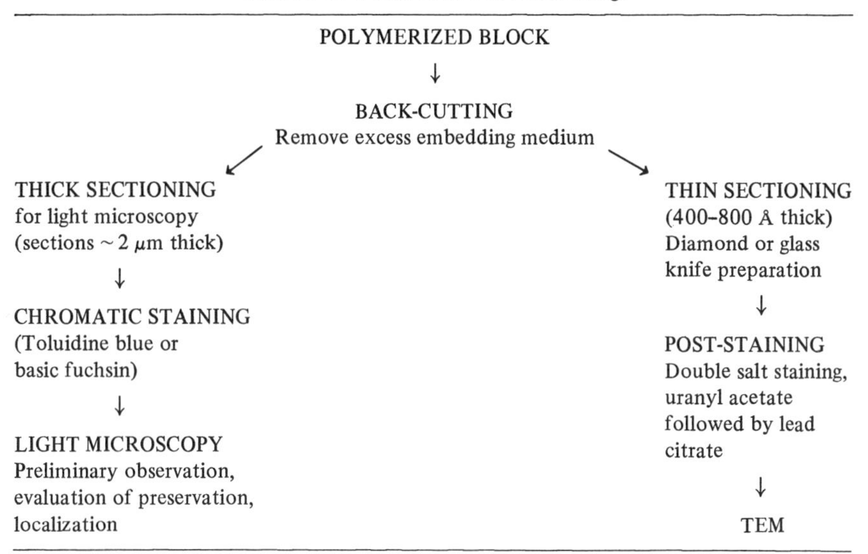

and the preparation of a trough (synonym: boat) for the flotation of thin sections. The diamond cutting edge is built with a trough; thus the user need not prepare this.

The trimmed specimen block and knife are mounted in an ultramicrotome and aligned, and the thin sectioning is begun. After a ribbon (i.e., a series of consecutively cut sections) is obtained, the tissue is mounted on a grid and post-stained to further enhance contrast. Commonly, the double salt staining technique of uranyl acetate followed by lead (e.g., lead citrate or acetate) is used. The prepared specimen is now ready for TEM examination.

To a large degree, success in thin sectioning presupposes that the specimen is properly embedded. That is, the specimen was thoroughly permeated with the correctly prepared embedding medium, and polymerization was complete. Tissues incompletely infiltrated tend to tear or pull out of the embedding matrix, and cannot be salvaged. A similar effect will be observed if the tissue was embedded but not infiltrated with accelerator-containing embedding medium; while the polymerized embedding medium may be properly cured, the tissue itself polymerizes more slowly. In the latter case it may be possible to continue polymerization, but severe extraction will be noted.

Table 8-2. Evaluation of a Properly Embedded Specimen.

PROBLEM	SOURCE	REMEDY
Tissue tears during sectioning or back-cutting	Insufficient infiltration	Increase duration of dehydration and infiltration
	Improper infiltration	Infiltration with accelerator-containing embedding medium
	High collagen content of tissue	Thin-section with a diamond knife
	Soft embedding content	Increase DDSA content of monomer
		Increase polymerization time or temperature
		Re-embed
Tissue shatters during sectioning or back-cutting	Reaction between reagents during processing	Glutaraldehyde and OsO_4—increase duration of intermediate buffer wash
		OsO_4 and ethanol—increase duration of intermediate water wash
	Characteristic of reagents themselves	Phosphate buffers and acetone both characteristically embrittle tissue; dehydrate with ethanol
	En bloc stain	Section with diamond knife or post-stain
	Excessive accelerator or hardener	Follow proper formulations
	Excessive polymerization	Follow proper temperature and time schedules

A too soft block also tends to tear rather than cleanly section specimens. The time and/or temperature of polymerization should be increased. This should be attempted before discarding the block, but if it is still soft after 1 day, discard it. This problem is due to improper embedding medium formulation; that is, the ratio of DDSA (the hardener) to the epoxy resin was too low, or insufficient accelerator was added. Various authors have re-embedded tissues when the tissue was irreplaceable (Ogura and Oda, 1973; McNelly and Hinds, 1975; Johnson, 1976).

On the other hand, tissues that have a high collagen content (e.g., skin) or dense inclusions (e.g., Reinke crystals in Leydig cells) are too dense for good sectioning with a glass knife. Consequently, a diamond knife should be used, or a glass knife if a minimal number of sections are to be cut.

At the opposite end of the spectrum are brittle blocks. This problem may be caused by an improper embedding medium formulation (i.e., excessive accelerator or hardener). Properly prepared media will become brittle if the temperature or duration of polymerization is overly high or long, respectively.

Embrittlement of the tissue (not necessarily of the embedding medium) may be a function of the processing reagents themselves or be due to reaction between the reagents. For example, phosphate buffers and acetone both increase tissue brittleness; to avoid this, use ethanol for dehydration. En bloc staining embrittles tissues, but when site-selective reactions are desired, the brittleness must simply be tolerated. An undesired reaction may occur between glutaraldehyde and osmium tetroxide if the buffer wash is insufficient; a similar reaction will occur between unreduced OsO_4 and ethanol if the intermediate water wash is incomplete. In either of these cases, one should increase either the duration of the wash or the number of fluid changes. Brittle blocks may be salvaged by sectioning with a diamond knife or a glass knife that has a small clearance angle. A summary of these problems, sources, and remedies is found in Table 8-2.

SPECIMEN BLOCK PREPARATION: BACK-CUTTING

The tip of a conventional BEEM capsules has an area of 1 mm^3, whereas most tissues are prepared as 0.5 mm^3 cubes. The excessive polymer surrounding the tissue should be removed for three reasons:

1. The tissue, not the embedding medium, is of interest.
2. It is difficult to cut large areas as ultrathin sections.
3. A proportionately larger area of the glass knife will be used and therefore dulled, whereas one prefers to have the maximum knife edge available for tissue sectioning.

To remove the tissue block from the BEEM capsule, first loosen the block by squeezing its sides with pliers, then push the tissue out by applying force at the tip of the capsule. The capsules are relatively tough if razor cutting is tried, and cutting of both the mold and one's fingers is quite common.

The tissue block is then mounted in a chuck or vise-type holder (e.g., Mack, 1964; Cheney and Ashurst, 1967; Leeper, 1968; Fineran, 1971), and examined at low magnification in a binocular light microscope. Various individuals use the ultramicrotome specimen holder chuck for back cutting, which is useful for positioning and simultaneously trimming the block in a good orientation (Scott and Thurston, 1975). The alternative method of trimming in the ultramicrotome will be considered below under "Ultramicrotomy"; Butler (1974) prepared a tool for rapid trimming that also works very well. Single-edged industrial razor blades are recommended; conventional shaving razors are too flexible. Several razors are cleaned with acetone, and should be replaced when the edge becomes dull. Using very short, shallow strokes, begin shaving away excess embedding medium. The goal here is to isolate the tissue as a shallow-angled pyramid topped by a trapezoid (Figure 8-1). The trapezoid will help in ribbon formation, but a square or rectangle is acceptable when serial sections are not required. The surface of the block should not be cut with a razor if the tissue is less than 0.5 mm deep; this is more easily done during fine trimming using the ultramicrotome. If more embedding medium separates the tissue from the block tip, carefully shave off the excess, but do not actually enter the tissue area. Again, the block face is fine-trimmed with the ultramicrotome and a glass knife.

A few hints on hand-trimming are as follows:

1. Always make very shallow, angled cuts when trimming the sides of the block; large cuts tend to make the specimen shatter.
2. Make the sides of the pyramid as smooth as possible; this helps in the formation of a straight ribbon during thin sectioning.

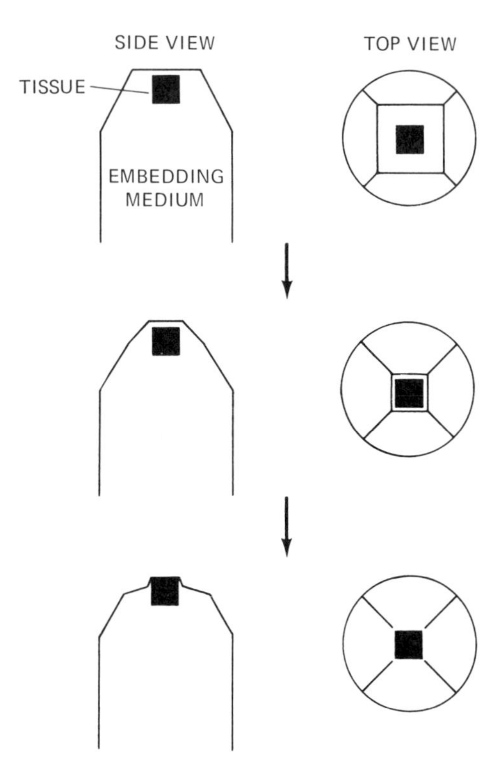

Figure 8-1. Trimming the specimen block.

3. Discard razor blades as soon as they show signs of dullness; a rough edge will impress itself on the tissue and gouge the specimen.

The better the specimen trimming, the easier the thin sectioning. Devices have been designed that help in back-cutting, all of which stress the above. Schneider and Sasaki (1976) developed a useful blade guide for shaky hands, and Guglielmotti (1976) and Gorycki (1978) also comment on fine-trimming methods.

KNIVES

Ultrathin sectioning requires that a cutting edge be extremely sharp and capable of producing sections less than 800 Å thick. The edge of a conventional metal knife used in light microscope histology is too inhomogeneous and dull for ultrathin sectioning (Gettner and

Ornstein, 1956) even though such knives were used in the 1940s for thin sectioning. Latta and Hartmann (1953) revolutionized thin sectioning when they discovered that a freshly cleaved piece of glass produced an extremely sharp knife edge. However, a glass knife edge rapidly becomes dull: Fernandez-Moran (1953, 1956) developed a knife composed of a single crystal of industrial diamond that maintained its sharpness over time and could thin-section specimens as dense as metals. As will be discussed, ultramicrotomes were also developed at that time and have been significantly modified since then, whereas glass and diamond knives still are the standard edges for ultrathin sectioning.

Glass Knives

The edge of a freshly cleaved piece of glass is literally the sharpest edge known to man, far surpassing any metal edge. Because of its nature as a supercooled liquid, however, the glass edge is subject to molecular flow, resulting over time in dullness. This is offset by the low cost of glass; knives are simply replaced when the edge becomes dull. For these same reasons, glass knives for ultramicrotomy are prepared immediately prior to use. Although prepared but old knives cannot be used for thin sectioning, they are appropriate for final back-cutting and thick sectioning with the ultramicrotome.

Glass suitable for knife making is commercially available (e.g., LKB or Dupont Sorvall) in strips approximately 1″ × 0.25″ × 16″. This glass is specially prepared for ultramicrotomy in that the strips have minimal stressmarks (e.g., hatchmarks or feathering on the long edges) and are very homogeneous; in comparison, glass purchased from, for example, a hardware store, is far less perfect and typically difficult to work with. With some skill it is possible to obtain two excellent knives per inch of LKB or Dupont glass, while one may need to consume a few inches of other glasses before producing an acceptable knife. The basic steps involved in preparing a glass knife are cleaning the glass, making two fractures, and finally attaching a waterproof trough for flotation of thin sections.

First, the strip of glass is cleaned with soap (preferably gritty soap) and thoroughly rinsed in hot water. The strip is then dried with a lint-free cloth if one is available; or it is air-dried. After this point, handle the glass only by one end (wrap a lint-free paper, e.g., lens paper, around this end). An acetone rinse will remove any trace contaminants, but be certain that the acetone itself is not oil-contaminated. Again air-dry the strip. When wiping the glass be careful not to cut yourself; the edges are extremely sharp. The glass must be scrupulously clean for the waterproof sealant used to attach a trough to the knife.

The fracturing of the glass strip is either by hand with modified glazier's pliers, or with commercially available knifemakers (e.g., Dupont/Sorvall, LKB, or Pelco Knifemaker). Knifemakers are standard equipment in a microscope lab, but the reader should be familiar with manual glass knife preparation.

The tools required for breaking by hand are modified glazier's pliers and a carbide scoring wheel, both of which are available from hardware stores. Weiner (1959) modified conventional glazier's pliers by placing a narrow strip (1–2 mm) of electrical tape (one or two layers thick) in the exact center of one jaw, and on the opposing jaw two identically sized strips of tape on both sides of and equidistant from the central strip. The single tape strip acts like a fulcrum, and the opposing strips are pressure points.

After cleaning the glass strip, measure 1″ from the end and make a straight score $\leqslant 0.5''$ in length in the center of the strip. The score must not extend from edge to edge because a free break during fracture is desired. The score must be at right angles to the side of the glass; a metal straightedge is conveniently used for delineating the score. Position the glass strip so that the score is just above the fulcrum (i.e., scored sides is in contact with jaw that has pressure points) and the pliers are at right angles to the entire strip of tape.

If a lengthy strip of glass is being prepared, prop up the long end so that the strip is parallel to the surface of the workbench. Slowly begin applying pressure by squeezing the pliers; this is difficult because the pressure application must steadily increase without fluctuation until the glass fractures. When the glass breaks properly, a moderate snap is heard; improper breakage is accompanied by a loud crack. The latter will induce stress within the glass square, making it difficult to obtain a good knife. Ward (1977a) clamped the pliers in a wood-working vise and was able to control the speed of the fracture much better than when the pliers were hand-held.

Secondly, the glass is rotated 45° and a second score ($\leqslant 0.5''$ long) is made a few degrees off center, but dividing the square into equal

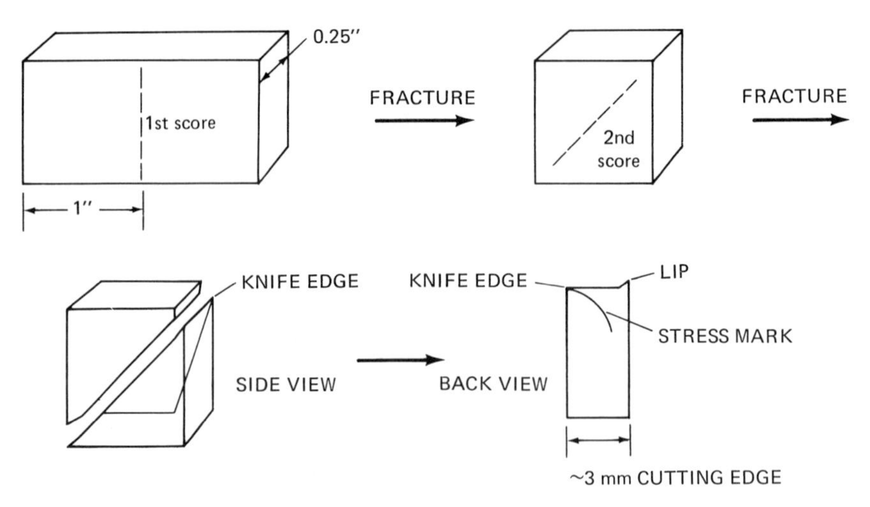

Figure 8-2. Preparation of glass knives.

portions (Figure 8-2). Grasp the square with the modified glazier's pliers, and slowly increase pressure until the glass fractures. Again, make sure that the score does not extend from edge to edge; the shorter score forces the glass to break freely (i.e., with minimal stress), resulting in the knife edge being sharper and having an angle greater than the scoring angle. Applying pressure too rapidly induces excessive stress in the glass, and the knives produced cannot be used for sectioning. Carefully lay the glass triangles on the workbench, and evaluate the knife edges, first with the naked eye and then with a binocular microscope (Figure 8-2; Sheldon, 1957; Squier and Randall, 1956). Discard knives that are very obviously curved. Frequently, knives are flat for one-half to two-thirds of their surface; then the remainder of the surface curves up and into a point (Andre, 1962). These knives are usable because, for example, if a 9-mm-wide knife has 4 or 5 mm of good cutting edge, this is quite large when compared to back-cut specimens with a maximum diameter of 0.5 mm. Simply do not use the curved area of the knife during sectioning, primarily because it is difficult to properly adjust the specimen-knife distance. Perhaps 10% of the time one will prepare a "perfect" glass knife, that is, an absolutely flat edge. In all glass knives, a prominent stress line originating at one corner and arcing down the back of the knife is visible. That corner of the knife must not be used for sectioning; it will mark the sections with deep stripes.

An alternative shape that results in a different knife angle is a rhombohedron. The same steps involved in preparing a 45° knife are used in preparing this ∿55° knife, as summarized in Figure 8-3. Any knife that has an acute angle is effectively sharper, and the effective clearance angle (to be discussed under "Ultramicrotomy") is larger. Ward (1977b) demonstrated that 55° knives may be prepared either by hand or with a knifemaker, and that this edge is indeed sharper than a conventional 45° knife.

A serious problem with this method is that it is very difficult to develop a "feel" for the proper ratio between speed and increase of force. Subsequently, a number of commercially available glass knifemakers have been developed that moderate the applied pressure. It is therefore possible to routinely produce good glass knives for thin sectioning. Knifemakers are also used for preparing large-diameter (∿12 mm) "Ralph" knives for light microtomy (Bennett et al., 1976); these are far too thick to fracture by hand. The individual should refer to the manufacturers' literature concerning glass-knife makers or Cosslett (1962), Fahrenback (1963), Fullager (1966), Griffin (1968), and Sutton (1969).

Glauert and Phillips (1967) determined that under optimal working conditions, up to 30 good EM sections may be cut on a given area of a glass knife; in more routine situations, perhaps 10 good sections may be cut. The used area of the knife is moved and sectioning started again until the entire usable edge is exhausted and another glass knife prepared. The cutting edge must be changed frequently because it dulls very rapidly; thin sections are so small that any im-

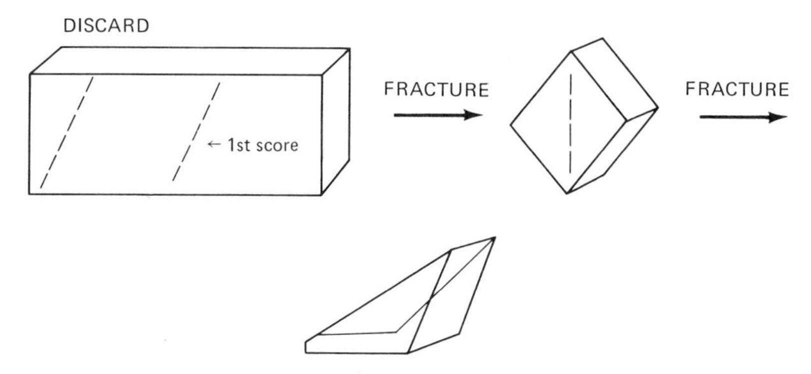

Figure 8-3. Preparation of a glass knife from a rhombohedron.

perfection (either native or due to wear) will damage the section. Roberts (1975) developed a technique that lengthens the lifespan of a glass knife edge. Using a conventional bell jar, tungsten is evaporated onto the knife edge; apparently the metal atoms become part of the edge, and if only a small amount is deposited, a compromise between metal and glass knives develops. This is useful if a diamond knife is unavailable, but the latter has the distinct advantage that it will cut many sections from the same specimen identically and without having to change operating conditions.

Ultrathin sections are extremely fragile and will adhere to a dry knife edge. This difficulty is overcome by surrounding the knife edge with a trough or boat to contain a flotation fluid (Hillier and Gettner, 1950). Although diamond knives are built directly into a trough, troughs must be prepared for glass knives. Various types of trough materials have been attempted, the most successful being metal, tape (e.g., electrical), or plastic troughs.

Nonreactive metal (e.g., stainless steel) boats are commercially available in various sizes. Gemmel and Henrikson (1970) introduced a large-capacity metal trough that is convenient for serial sectioning of autoradiographic specimens for electron microscopy. The advantages of metal troughs are that they are inert, readily cleaned, and reusable.

One positions the metal trough on the glass knife, taking care to ensure that the upper edge of the trough is at or above the level of the knife. McKinney (1969) describes an ingenious method for waterproofing and sealing the trough to a knife. A small piece of dental wax is placed in the lowest point of the trough and melted by pointed heat application with an alcohol dental torch. The sides of the trough are gently heated, and capillary action draws the melted wax up the sides. After a minute or so of cooling, the wax will solidify and form a waterproof seal. Provided that the glass is clean, this very thin strip of wax is a very efficient sealant. If traces of oil are on the glass, no amount of wax will seal the trough to the knife. The troughs are cleaned in acetone or xylene to remove excess wax.

An alternative material for preparing troughs is tape. Electrical tape works very well; its black color makes it easy to distinguish interference colors of thin sections without glare. To prepare a tape trough, first place a strip of electrical tape (2 X 8 cm) on a clean glass plate. Using a razor and a metal straightedge, cut the tape into four sections with dimensions of ∿1 X 4 cm. Each section may be used for boats. Remove one strip of tape, and holding it by only one end, align the tape so that it is parallel to the bottom of the glass knife, at or above the knife edge, and its width is parallel to the back of the knife (Figure 8-4). (Note: Placing the tape at a 2–3° angle relative to the bottom of the glass is acceptable; this compensates for tilting the knife during ultramicrotomy.) Press the tape against the glass, and wrap it around the angled portion of the knife. Press on the opposite side of the edge, again making sure that the top of the tape is at or above the level of the knife edge. Remove the excess tape by cutting with a sharp razor and taking care not to touch the knife edge. The bottom of the tape, especially the gap at its lowest point, is then sealed with a waterproof material. Fernandez-Diez and Gonzales-Angulo (1970) recommended nail polish as a sealant, but it requires longer drying times than dental wax. Sheet dental wax, a metal spatula, and an alcohol burner are required as follows: Heat the spatula for a few seconds, and press it under the wax until a small amount melts. Immediately draw the sharp edge of the spatula around the trough, putting an extra drop at the gap between the tape and knife. This forms a very thin but most effective sealant; again, excess wax will not seal dirty glass. Other ways to apply wax include painting with molten bee's wax–but this is messy, bee's wax fumes are carcinogenic, and one has a tendency to put an excess on the boat. Especially along the sides of the knife, too much wax may interfere with sectioning. Always reexamine the knife edge after these operations to ensure that it has not been damaged. One prob-

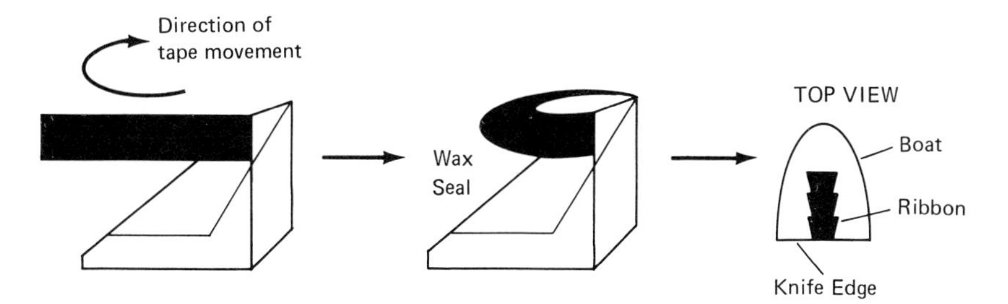

Figure 8-4. Preparation of a tape trough.

lem with tape troughs is that the inside of the trough is, of course, sticky, and may contaminate the flotation fluid.

Gulati and Akers (1977) prepared nonreactive, disposable troughs using commercially available plastic microslides. The slides or portions thereof are shaped and the edges cut with a razor, positioned around the knife edge much like a metal trough, and sealed with wax. Various sizes of troughs are thus easily prepared, and they will not react with, for example, water when it is used as the flotation fluid. On the other hand, the EM novice may experience some difficulty seeing the events occurring at the surface of the flotation fluid because these troughs are opaque.

Usually it is most convenient to prepare two knives with troughs before beginning sectioning, so that if the first knife becomes dull, a substitute is immediately available. If the second is not used, it may be saved for thick sectioning. In any event, glass knives for cutting $\sim$2-μm sections rarely need attached boats; in most situations thick sections are sufficiently sturdy to be cut on a dry knife.

Diamond Knives

Situations exist in which a glass knife simply is not appropriate. For example, cartilage has a high collagen content and resists cutting with glass; the knife will become dull after cutting very few sections. Also, only a limited number of sections of any type of tissue can be cut before a glass knife becomes dull, meaning that serial sectioning a few microns' worth of tissue is next to impossible. In these cases, a diamond knife is a much more suitable cutting edge. Developed by Fernandez-Moran (1953, 1956), diamond knives have the advantages of consistently and easily cutting high-quality sections of reproducible thickness, the sections are thinner than those obtained with a glass knife, they are used repeatedly provided they are properly cleaned after use, and virtually no preparation of the knife is necessary (i.e., the diamond chip is permanently mounted in a metal trough). Diamond knives are also very expensive ($\sim$\$1500.00, depending upon the usable length of the cutting edge), require a great deal of care in handling (EM novices do not learn to thin-section with diamonds), and periodically during normal use require resharpening. The care of a diamond knife is critical because if, for example, the cutting edge is nicked, the damage will remain forever; diamonds may be resharpened after normal use, but they cannot be mended if damage is severe.

The cutting edge of a diamond knife is commercially available in lengths of 1.0–6.0 mm, with cost proportionate to the length. The typical length for routine applications is 1–3 mm, with the knife situated in the boat at an angle of 40–48°; the angle may be preset up to 60° by the manufacturer for cutting extremely hard specimens such as metal. Depending upon the manufacturer, industrial diamonds are cleaved along natural planes, or single-crystal alluvial diamonds are used. Abrasive grinding and polishing enhance the natural cutting properties of the diamond. The prospective knife is then permanently mounted in a boat and evaluated by both light microscopy and actual ultrathin sectioning.

The proper care of a diamond knife involves cleaning the edge immediately after use and while it is still wet. Pithwood or balsa sticks are flattened at one end, dipped in pure acetone, and gently drawn along the length of the knife, parallel to the edge. Be very careful to parallel the knife edge; any other orientation will damage the diamond. Gorycki and Oberc (1978) caution that this cleaning method may damage the knife unless extreme care is taken at the stick/diamond orientation level; alternately, they recommend that a solution of 7x (a cleaning agent) in distilled water be prepared and filtered through an 0.8-μm Nalgene filter. The diamond edge is carefully submerged in the soap for $\sim$10 min., and the edge is squeegeed with 1.6-mm-diameter Tygon tubing impaled on a needle. This same soap can be used for cleaning the trough followed by rinsing with distilled water and drying. Ultrasonic cleaning is another method (Wallstrom and Iseri, 1972). Carefully remove excess flotation fluid, rinse the inside with water, and dry the boat, and finally store the glass knife in a container supplied by the manufacturer. With careful handling, a diamond knife can be used for years. The edge may periodically require resharpening, which is done by the manufacturer at a cost of approximately one-half the original value of the knife. For this reason, many laboratories keep two diamond knives on hand; the second is used while the other is resharpened.

When diamond knives are used for ultramicrotomy, it is recommended that a glass knife be used for fine trimming the block face in the microtome. Although the knife angle may be different, fine trimming and/or thick sectioning is never done with a diamond. Sections thicker than ∿1 μm will damage the diamond; this corresponds to an interference color of greenish-blue. Use a dry glass knife (i.e., one without a trough) and trim the specimen; then insert and align the diamond knife for thin sectioning. See Chaplin (1972) and Heuser (1980) for a method of diamond knife/specimen alignment using reflecting foils.

ULTRAMICROTOMY

The ultramicrotome is a specialized instrument capable of cutting ultrathin sections in the range of 400–800 Å thickness; that is, a thickness that can be penetrated by the electron beam. Although early electron microscopists used conventional microtomes such as those used in light histology, it readily became apparent that an instrument capable of very small advances per stroke was necessary. Concurrent with the development of glass and diamond knives, various researchers introduced ultramicrotomes for electron microscopy that provided the basis for all commercially available instruments. In 1953 Porter and Blum introduced a mechanical-advance ultramicrotome that has evolved into the sophisticated series of automatic ultramicrotomes marketed by Sorvall (e.g., Sorvall MT-1, MT-2B, and MT-5000 systems). Concurrently, Sjostrand (1953) developed an ultramicrotome that was modified by Hellstrom (1960), evolving into the LKB Ultratomes (e.g., LKB UM III, IV). The Cambridge Huxley Mark 1 and 2 were invented by Huxley (1957) and are marketed by LKB (except in Great Britain where they are available from Kent Cambridge Medical). Sitte (1953) developed a thermal-advance instrument, which has grown into the Reichert Ultramicrotomes (OMU2, OMU3). The following discussion is designed to be appropriate for use of any of these ultramicrotomes; Reid (1975) presents an excellent overview of these systems, or the reader may contact the individual manufacturers for detailed literature.

Essentially, an ultramicrotome functions in the following manner: the specimen arm, holding a sample, will travel at a constant speed past the fixed knife edge, retract, and repeat the cycle, advancing X nm per stroke. When the distance separating the specimen from the knife edge is less than the distance advanced per stroke, a section will be cut. Repetition of this cycle (advance, cut, retract, advance . . .) will cause a ribbon of sections to form; the individual sections composing the ribbon should be of the same thickness.

The advance system of the ultramicrotome serves a number of purposes. The coarse advance is used for initially aligning the knife relative to the specimen, and the fine advance ($\geqslant$0.25 μm) is used for thick sectioning and block trimming. Both coarse and fine advances move the knife toward the specimen. In comparison, the ultrafine advance moves the specimen toward the knife in distances $\leqslant$1000 Å per stroke. Whereas the coarse and fine movements are directly manipulated by the microtomist, the ultrafine advance is regulated by the instrument itself; the microtomist selects the desired degree of advance (e.g., 800 Å), and the ultramicrotome is responsible for moving 800 Å forward per stroke. This linear advance is thermal or mechanical in origin.

Some of the factors that adversely affect this stroke are friction between the specimen and knife, the generation of heat or static electricity from the specimen rubbing on the back of the knife, vibrations and room drafts. Friction and heat are controlled by the clearance angle of the knife (to be discussed) and using a thermally conductive flotation fluid. To minimize vibrations, ultramicrotomes are normally positioned on rubber absorbing pads or special tables, and assembled in isolated rooms during installation. The microtomist should take care not to personally introduce vibrations or drafts; do not lean on the microtome bench, and lock yourself into the laboratory so that you will not be disturbed by doorway drafts. Static electricity is sometimes a problem, especially when ambient humidity is very low; Nicholson (1978) designed a corona electrode discharge probe that may be attached to the ultramicrotome to relieve this problem.

The rate at which the specimen passes the knife edge is referred to as the cutting speed. Successful sectioning requires that the cutting speed be constant; if it is varied while the specimen is in contact with the knife, sections are of very low quality. This is one of the major reasons why automatic ultramicrotomes are preferred over manual systems; the early Porter-Blum models, for example, required that the microtomist control the cutting speed by hand, which is a difficult maneuver. All modern ultramicrotomes have variable cutting speeds. A rate of 2–3 mm/sec is generally applicable for specimens of moderate density (e.g., most tissues), whereas the rate is slower or faster for a hard or a soft specimen, respectively.

Let us now consider the real-time event of section production. Compression occurs with separation of a thin section from the bulk tissue when it is forced through a sharp edge, the degree of compression being a function of knife sharpness. A very sharp knife will readily pass through the tissue with minimal compression, whereas a dull knife resists passing through the specimen and increases compression; these situations produce, respectively, a good and a poor section.

In addition to the degree of compression, the clearance angle of the knife relative to the specimen face affects the section quality (Figure 8-5). The angle should be large enough that the specimen clears the back of the knife after striking its edge, but small enough to maintain knife sharpness. This angle is normally between 2° and 4° (as set on the ultramicrotome), but recall that glass knives prepared from rhombi inherently have a larger angle. This factor should be taken into account if the real clearance angle is to be known. As a rule of thumb, excessively large clearance angles (>5°) rapidly dull the knife edge (Moretti et al., 1970).

In isolated situations, for example, those in which sections show too much compression even with a new glass knife, or a different clearance angle or varied cutting speed was unsuccessful, thin sections may be artificially expanded. Dipping a cotton swab in chloro-

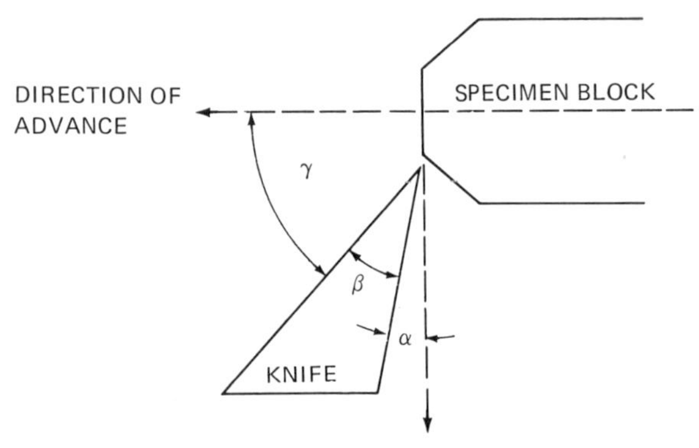

Figure 8-5. Specimen-knife geometry. α = clearance angle, β = knife angle, γ = sheer angle.

form and passing it over the sections (chloroform fumes only, not the liquid) will expand the sections (Sotello, 1957; Satir and Peachey, 1958). Commercially available heat pens may also be used, and have the advantage that the degree of expansion is easily controlled (Roberts, 1970).

All of these sectioning events are occurring on the surface of a liquid-filled trough. Thin sections will only be cut if the knife edge is wet; the fragile sections will crumple on a dry edge. Thus, the trough is filled with a flotation fluid, which serves to wet the knife edge, as a substrate for thin sections, and to accommodate the heat/static electricity generated during sectioning (Gettner and Ornstein, 1956). Distilled water is usually used as a flotation fluid for both glass and diamond knives. A new diamond knife, however, is slightly hydrophobic, although it loses this characteristic over time (Porter, 1964). If excessive hydrophobicity is present, 10% acetone in water, which has a lower surface tension than pure water, may be used.

The trough is filled with the flotation fluid only after the knife and specimen have been aligned in the ultramicrotome. A syringe or finely drawn pipette is filled with the fluid and added to the trough until a slightly positive meniscus forms; this guarantees that the knife edge is wet. Then the excess is drawn off with a torn filter paper until a flat meniscus is observed. Take care that the liquid does not run out over the knife edge (recall that the knife is tilted); if the back of the knife is wet, sections will not be cut, and the block face will be wet as it is drawn past the knife. On the other hand, if a negative meniscus exists, the sections will crumple at the knife edge. The flotation fluid will seep out if the boat is not waterproof; thus the importance of using clean glass becomes clear. The level of the fluid will decrease over time, especially after thin sections are removed, and it should be replenished until it is again flat. Many beginners forget this, and become frustrated when they do not produce thin sections. Simply make a habit of routinely checking the flotation fluid level.

METHOD OF THIN SECTIONING

Following the back-cutting of a specimen block, and the preparation of one glass knife without a trough and one or two glass knives with troughs, one is ready to use the ultramicrotome. The additional necessary materials, which should be at hand before beginning, are acetone-cleaned grids (200- or 300-mesh), watchmaker forceps, a few sheets of filter or lens paper, a syringe, the flotation fluid, and a single-hair brush. The latter is simply an eyelash or small camel's hair taped to the end of an applicator stick. It is very useful for manipulating sections in the trough and removing extraneous tissue fragments. Because the hair may adhere to sections, Teflon probes may be easily prepared by cutting them from a Teflon foil (Reichle, 1972; Nyhlen, 1975). The Teflon does not damage the section if only its edge is touched. When thick sectioning for light microscopy is desired, add microscope slides to the list. The procedures for chromatic staining and preparing permanent mounts are discussed in Chapter 10. This very basic procedure is touched upon here because this is the usual point at which it occurs.

It should be noted that special instruments are available for the following operations that precede thin sectioning. However, few labs are sufficiently wealthy to have such equipment.

Regardless of the type of ultramicrotome used, the following procedures apply. First, the specimen block is fine-trimmed with a glass knife. Mount the block in the specimen holder with the long edge of the trapezoid facing down. This edge will strike the knife first and, because it is large, reduces compression. Push the capsule into the collet jaw until only a few millimeters extend outward. If more than 5 mm of the block is visible, remove it and file the end (opposite the specimen) to remove excess medium. Firmly lock the specimen into place in order to reduce the probability of chatter (Glauert and Phillips, 1967), which is a very regularly spaced variation of thickness within a section caused by vibration (Phillips, 1962). Flat castings are mounted in manufactured chucks in much the same manner (Nelson and Flaxman, 1972; Dalen, 1973; Godkin and Keith, 1975).

Second, retract the knife advance mechanism and insert the plain glass knife perpendicular to the specimen plane. Situate the knife so that it is flush with the back of the knife-holding assembly, and adjust its height according to manufacturer specifications. Most ultramicrotomes simply have an indicator for the proper height, which corresonds to the area where the cutting speed is regulated. Manually move the specimen arm toward the midpoint of its cutting stroke which is just above the level of the knife edge. Manually cycle the microtome a few times with the knife retracted to ensure that you become familiar with the stroke; for example, Sorvall microtomes retract the specimen back and then it moves forward, whereas Huxley microtomes cycle through a parallelogram.

Return the specimen to knife level; from this point on, all observations are made with the attached binocular microscope focused at the knife edge. Using the coarse advance, move the knife toward the sample until the distance separating the two is $\leqslant$1 mm. Switching to the fine advance of the knife, move toward the specimen $\sim$2 μm per microtome cycle (i.e., manually move the specimen with, e.g., the flywheel of Sorval microtomes while simultaneously using your other hand to advance the knife). This can be done quite rapidly until sections are cut; then reduce speed. The excess embedding medium will be finely shaved from the tip of the specimen. As soon as contact with the tissue occurs (the tissue is very dark in the lighter embedding medium), stop, retract the knife, and move to another good area of the knife. Readvance and carefully cut thick sections until all excess medium is removed. If any regular scratches are visible on the sections, it is essential that a different area of the knife be used; these scratches gouge the surface of the specimen.

Sections 2–3 μm thick may be prepared now (Galey, 1963). Slowly carry the specimen past the knife and observe the thick section adhearing to the knife edge; it may be flat if very thick (3–4 μm) or curled if thinner (2–3 μm). Either thickness can be used. Carefully pick up the section by its edge with fine-tipped forceps or by placing a small wood splinter or single-hair brush beneath the section, and transferring it onto a drop of water on a clean microscope slide. Gently heat the slide over an alcohol burner. Curled sections will flatten out as the water evaporates and the sections heat-fix to the slide. Chromatic staining is easily done by placing a drop of toluidine blue O on the section; gently heat it for a few seconds, and observe the tissue uptake. As soon as the section becomes bluish (5–30 sec), rinse with distilled water, heat-dry, and examine it. A permanent mount may be prepared by placing a drop of embedding medium or conventional coverslipping medium and a glass coverslip on this section. Observe the thick section with a light microscope. The block is now ready for thin sectioning with a diamond or glass knife.

The earlier discussion of specimen block preparation mentioned that it is possible to do all of the trimming with the ultramicrotome and a glass knife. The beginner will find it simpler to use razors for this step, but with some experience it is quite easy to back-cut with a microtome. Follow all of the steps given above for mounting the knife and block, beginning with tip trimming. When the tissue is reached, rotate the knife 25–30° to either side of the block edge and cut until the tissue is reached, and continue rotating the knife/specimen until all sides of the tissue have been cut into the trapezoidal shape. It is simpler to avoid the pyramid shape (i.e., mimic the shape of the capsule). Alternatively, the so-called mesa technique may be modified. Originally developed by DeBruijn and McGee-Russell (1966), the mesa technique is used for isolating specific areas of interest within a specimen (e.g., a mesa consisting solely of glomerulus rather than other areas of kidney), which saves time when one is searching for a certain structure in the TEM. It simply involves removal of excess tissue by thick sectioning around the area of interest, resulting in a small raised area at the tip of the block. The specimen is rotated and the knife moved laterally for cutting.

The specimen is now ready for thin sectioning. First, retract the knife mechanism and remove the used glass knife. Do not change the specimen orientation, and proceed as follows:

1. Reset the automatic advance system to ensure that maximum travel is available, and the cutting speed to $\sim$3 mm/sec.
2. Set section thickness to its maximum value; this will be reduced as soon as sections are cut.
3. Place the glass knife with trough in the instrument flush with the back of the knife-holding assembly, adjust height, and secure it in place. Set the clearance angle to 3°; this may need changing if the block is soft or brittle. Evaluate the knife edge and laterally move it until a good cutting area is roughly parallel to the tip of the specimen.
4. Manually cycle the ultramicrotome until the specimen is at the midpoint of its real sectioning range (knife edge plane).
5. Manually move the knife toward the specimen and check that the lower edge of the specimen is parallel to the knife edge. This is essential for straight ribbon formation.
6. Move the light source until a reflection is seen originating from

the block face. Advance the knife to the point where it is reflected by the block face (Gorycki, 1975, 1978). This will also reveal if the block edge is exactly parallel to the knife edge; if it is not, retract the knife and rotate the specimen. It is not recommended that the knife be moved; different stress problems, all negative, develop with angled sectioning.

7. Continue advancing the knife until a minute distance separates the real knife edge from its reflection; then stop. This is very easily done with point light sources or fiber optics, which are now standard on state-of-the-art instruments (Gartner, 1965). For this and other methods of alignment, see Gartner (1965), Kindel (1970), Boere (1971), Nelson and Flaxman (1973a,b), Nickels (1974, 1975) Wyatt (1974), Gorycki (1975, 1978) Isler (1975), Butler (1976), and Mollenhauer (1976).
8. Manually cycle the microtome through three-quarters of a cycle (i.e., until the specimen is at its uppermost limit). Fill the trough with the flotation fluid until it has a slightly positive meniscus; then draw off the fluid until the meniscus is flat. Focus the binocular microscope on the knife edge. Move the light source until the fluid surface near the knife edge is reflective. This is easiest if the binoculars are tilted toward the operator. Section thickness is estimated by observing interference colors, but they are seen only when the above conditions are met.
9. Place the ultramicrotome in the automatic mode, and let it cycle at maximum thickness until a section is cut. Immediately evaluate its thickness by observing its interference color. Using ellipsometry, Peachey (1958, 1960) devised a chart correlating section color with approximate thickness:

Interference color	∿ Thickness
Gray	60 nm
Silver	90
Gold	150
Purple	190
Blue	240

For routine purposes, this chart is very useful. Various persons have shown, however, that the dimensions noted above may vary significantly within a given solid color section as well as among sections of the same color (e.g., Mota, 1960; Zelander and Ekholm, 1960; Phillips and Shortt, 1964; Williams and Meek, 1966; Cosslett, 1967; Helander, 1969; Silverman et al., 1969; Weibull, 1970; Gillis and Wibo, 1971; Willis, 1971).

10. Decrease the section thickness with the ultramicrotome's thickness control until a silver-gold interference color is observed. These sections are usually used, although high-resolution microscopy requires sections 60–90 nm thick.
11. A ribbon (series) of sections should be cut, all of which have the same interference color. Stop the advance mechanism, and gently push the sections away from the knife edge with the

Table 8-3. Problems in thin sectioning.

PROBLEM	SOURCE	REMEDY
No sections are cut	Wet block face and/or back of the knife	Dry each with lens paper
	Static charge	Pat the block face with acetone-soaked lens paper
	Advance system not reset	Reset to maximum travel
Sections exhibit uneven thickness	Soft block	Increase polymerization duration
	Dull knife	Replace knife
	Loose specimen or knife	Check clamps
	Chatter at TEM level	Insert block deeper into specimen arm, or change the clearance angle
	Vibration	Avoid excess physical contact with ultramicrotome
	Cutting speed	Decrease rate
Poor ribbon formation	Block face upper and lower edges are not parallel	Back-cut more carefully, or align block face parallel to knife edge
	Sections leave knife edge	Lower meniscus level
Scratches visible	Particle embedded in knife edge	Move knife into another cutting area
Sections crumple	Poor embedding	Increase polymerization duration
	Negative meniscus	Increase fluid level
	Dull knife	Replace knife
Sections fly away	Static charge	Blot block face with acetone/lens paper; use corona discharge probe

single-hair brush. Grasp a clean grid with forceps and submerge it in the boat. Slowly bring it toward the ribbon, position the grid at a slight angle beneath the ribbon, and pull it directly upward. The tissue will adhere to the matte side of the grid; gently blot the shiny side until the grid is dry. The tissue is now ready for post-staining and TEM examination. Table 8-3 summarizes more common problems and remedies for thin sectioning.

REFERENCES

Andre, J. (1962) Presentation d'un appareil destine a preparer les couteaux de verre. *J. Ultrastr. Res.* 6:437.

Bennett, H. S., A. D. Wyrick, S. W. Lee, and J. H. McNeil. (1976) Science and art in preparing tissues embedded in plastic for light microscopy, with special reference to glycol methacrylate, glass knives, and simple stains. *Stain Technol.* 51:71.

Boere, H. (1971) Methods for orienting and aligning the pyramid front-face and knife-edge on the LKB Ultratome ultramicrotomes I and II. *Sci. Tools* 18:24.

Butler, J. K. (1974) A precision hand trimmer for electron microscope tissue blocks. *Stain Technol.* 49:129.

—— (1976) An illuminator for ultramicrotome knife orientation and block approach. *Stain Technol.* 51:241.

Chaplin, G. L. (1972) Facilitating diamond knife ultramicrotomy by the use of reflecting metallic foil. *Stain Technol.* 47:47.

Cheney, R. A., and D. E. Ashurst. (1967) A method for the orientation of tissues in epoxy resin-blocks, and a design for a rotation stage for holding blocks during trimming. *J. Roy. Micros. Soc.* 86:441.

Cosslett, A. (1962) Preparation of glass knives for thin section cutting. *J. Roy. Micros. Soc.* 80:301.

—— (1967) Letters to the editors. *J. Roy Micros. Soc.* 86:315.

Dalen, H. (1973) Vertical sectioning. A. Millipore filters. In: *Tissue Culture, Methods and Applications.* Kruse, P. F., and M. K. Paterson (eds.). Academic Press, New York.

DeBruijn, W. C., and S. M. McGee-Russell. (1966) Bridging a gap in pathology and histology. *J. Roy. Micros. Soc.* 85:77.

Fahrenbach, W. H. (1963) A contribution to glass knife breaking. *J. Cell Biol.* 18:475.

Fernandez-Diez, J., and A. Gonzalez-Angulo. (1970) Sealing troughs to glass knives: Advantages of fingernail lacquer (nail polish). *Stain Technol.* 45:190.

Fernandez-Moran, H. (1953) A diamond knife for ultrathin sectioning. *Exp. Cell Res.* 5:255.

—— (1956) Applications of a diamond knife for ultrathin sectioning to the study of the fine structure of biological tissues and metals. *J. Biophys. Biochem. Cytol., (Suppl)* 2:29.

Fineran, B. A. (1971) A device for holding LKB Ultratome chucks during preliminary block trimming. *J. Micros.* 94:83.

Fullager, K. (1966) The role of the LKB Knifemaker in ultramicrotomy. *Sci. Tools* 13:39.

Galey, F. (1963) A mechanical technique for trimming tissue blocks in electron microscopy. *J. Ultrastr. Res.* 9:139.

Gartner, J. (1965) The slit lamp as an aid in ultramicrotomy. *Sci. Tools* 12:43.

Gemmel, R. T., and R. C. Henrikson. (1970) A large trough for mounting thin sections on slides. *Stain Technol.* 45:135.

Gettner, M. E., and L. Ornstein. (1956) Microtome. In: *Physical Techniques in Biological Research*, vol. 3. Oster, G., and A. W. Pollister (eds.). Academic Press, New York.

Gillis, J. M., and M. Wibo. (1971) Accurate measurement of the thickness of ultrathin sections by interference microscopy. *J. Cell Biol.* 49:947.

Glauert, A. M., and R. Phillips. (1967) The preparation of thin sections. In: *Techniques for Electron Microscopy*, 2nd ed, p. 213. Kay, D. H. (ed.). Blackwell Scientific Pub., Oxford.

Godkin, S. E., and C. T. Keith. (1975) Ultramicrotome chucks for flat castings. *Stain Technol.* 50:63.

Gorycki, M. A. (1975) Aligning block faces by reflected light for precise sectioning. *Stain Technol.* 40:265.

—— (1975) Simple and rapid block face alignment methods for the ultramicrotome. *Stain Technol.* 52:255.

—— (1978) Methods for precisely trimming block faces for ultramicrotomy. *Stain Technol.* 53:63.

——, and M. A. Oberc. (1978) Cleaning diamond knives before or during sectioning. *Stain Technol.* 53:51.

Griffin, R. L. (1968) Machine-made glass knives in electron microscopy. *J. Med. Lab. Technol.* 25:146.

Guglielmotti, V. (1976) Device for manual trimming of tissue blocks for ultramicrotomy. *Stain Technol.* 51:135.

Gulati, D. K., and S. W. Akers. (1977) Improved trough for ultramicrotomy. *Stain Technol.* 52.351.

Helander, H. F. (1969) Surface topography of ultramicrotome sections. *J. Ultrastr. Res.* 29:373.

Hellstrom, B. (1960) The Ultratome Ultramicrotome-basic principles and summarized description of construction. *Sci. Tools* 7:10.

Heuser, R. C. (1980) A rapid diamond knife alignment procedure. *Proc. 38th Ann. EMSA Meet.*, p. 648.

Hillier, J., and M. E. Gettner. (1950) Improved sectioning of tissues for electron microscopy. *J. Appl. Phys.* 21:889.

Huxley, A. F. (1957) An ultramicrotome. *J. Physiol.* 137:73.

Isler, H. (1975) Simple mechanical device for orienting tissue blocks on the ultramicrotome. *J. Micros.* 102:225.

Johnson, P. C. (1976) A rapidly setting glue for resectioning and remounting epoxy embedded tissue. *Stain Technol.* 51:275.

Kindel, L. (1970) Rational preparation prior to ultrathin sectioning. *Proc. 7th Int. Cong. EM* 1:419.

Latta, H., and J. F. Hartmann. (1950) Use of a glass edge in thin sectioning for electron microscopy. *Proc. Soc. Exp. Biol. Med.* 74:436.

Leeper, G. F. (1968) A transilluminating holder for epoxy blocks. *Stain Technol.* 43:291.

Mack, J. P. (1964) A holder for trimming tissue blocks for electron microscopy. *Stain Technol.* 39:177.

McKinney, R. V. (1969) Facilitation of sealing metal troughs to glass knives by use of an alcohol hand torch and dental baseplate wax. *Stain Technol.* 44:44.

McNelly, N. A., and J. W. Hinds. (1975) Rescuing poorly embedded tissue for electron microscopy: A new and simple technique for re-embedding. *Stain Technol.* 50:209.

Mollenhauer, H. H. (1976) Improved specimen lighting in ultramicrotomy by painting reflective surfaces on specimen blocks. *J. Micros.* 107:203.

Moretti, G. F., U. Barbolini, and A. Baroni. (1970) A method to adjust a glass knife in the LKB Ultrotome to a suitable clearance angle. *Sci. Tools* 17:52.

Mota, M. (1960) Sectioning of sections for determination of thickness in electron microscopy. *Melhoramento* 13:127.

Nelson, B. K., and B. A. Flaxman. (1972) In situ embedding and vertical sectioning for electron microscopy of tissue cultures grown on plastic petri dishes. *Stain Technol.* 47:261.

—— and B. A. Flaxman. (1973a) Monitoring of epoxy block trimming by the use of a laterally mounted mirror. *Stain Technol.* 48:9.

—— and B. A. Flaxman. (1973b) Use of high intensity illumination to aid alignment of knife edge and block face for ultramicrotomy. *Stain Technol.* 48:13.

Nicholson, P. W. (1978) A device for static elimination in ultramicrotomy. *Stain Technol.* 53:237.

Nickels, J. (1974) Fiber optics in the illumination of epoxy resin embedded cells. *Micros. Acta* 76:48.

—— (1975) An ultramicrotome illuminator for epoxy resin embedded single cell specimens. *Stain Technol.* 50:359.

Nyhlen, L. (1975) A modified method for preparing teflon-tipped probes for manipulation of thin sections. *Stain Technol.* 50:365.

Ogura, H., and T. Oda. (1973) A method for the recovery of inadequately epoxy resin–embedded tissues for electron microscopy. *J. Electron Micros.* 22:365.

Peachey, L. D. (1958) Thin sections. I. A study of section thickness and physical distortion produced during microtomy. *J. Biophys. Biochem. Cytol.* 4:233.

—— (1960) Section thickness and compression. *Proc. 4th Int. Cong. EM*, p. 2P-72.

Phillips, R. (1962) Comment on localized, short-spaced periodic variation in contrast on methacrylate sections. *J. Roy. Micros. Soc.* 81:41.

—— and T. Shortt. (1964) The "re-sectioned section" technique and its application to studies of the topography and thickness of thin sections. *J. Roy. Micros. Soc.* 82:263.

Porter, K. R. (1964) Ultramicrotomy. In: *Modern Developments in Electron Microscopy*, C-4. Siegel, B.M. (ed.). Academic Press, New York.

—— and J. Blum. (1953) A study in microtomy for electron microscopy. *Anat. Rec.* 117:685.

Reichle, R. (1972) A teflon-tipped probe for easy manipulation of ultrathin sections in the knife trough. *Stain Technol.* 47:171.

Reid, N. (1975) Ultramicrotomy. In: *Practical Methods in Electron Microscopy*, vol. 3 part 2. Glauert, A. M. (ed.). American Elsevier, New York.

Roberts, I. M. (1970) Reduction of compression artifacts in ultrathin sections by the application of heat. *J. Micros.* 92:57.

—— (1975) Tungsten coating–a method of improving glass microtome knives for cutting ultrathin sections. *J. Micros.* 103:113.

Satir, P. G., and L. D. Peachey. (1958) Thin sections. II. A simple method for reducing compression artifacts. *J. Biophys. Biochem. Cytol.* 4:345.

Schneider, H., and P. J. Sasaki. (1976) A blade guide for hand trimming resin blocks for ultramicrotomy. *Stain Technol.* 51:283.

Scott, J. R., and E. L. Thurston. (1975) Ultramicrotome specimen chuck modification for improved viewing in block trimming. *Stain Technol.* 50:290.

Sheldon, H. (1957) A method for evaluating glass knives. *J. Biophys. Biochem. Cytol.* 3:621.

Silverman, L. B. Schreiner, and D. Glick. (1969) Measurement of thickness within sections by quantitative electron microscopy. *J. Cell Biol.* 40:768.

Sitte, V. (1953) Ein einfaches ultramikrotom fur hochauflosende elektronenmikroskopische untersuchungen. *Mikroskopie* 10:365.

Sjostrand, F. S. (1953) A new microtome for ultrathin sectioning for high resolution electron microscopy. *Experientia* 9:114.

Sotello, J. R. (1957) Technical improvements in specimen preparation for electron microscopy. *Exp. Cell Res.* 13:599.

Squier, C. A., and M. Randall. (1965) A simple adjustable holder for the inspection of glass knives. *J. Roy. Micros. Soc.* 84:571.

Sutton, J. S. (1969) Producing improved glass knives for ultramicrotomy; a glass breaker featuring a linear fulcrum and a device for controlling fracturing velocity. *Stain Technol.* 44:287.

Wallstrom, A. C., and O. A. Iseri. (1972) Ultrasonic cleaning of diamond knives. *J. Ultrastr. Res.* 41:561.

Ward, R. T. (1977a) A method for breaking glass knives slowly. *Stain Technol.* 55:116.

—— (1977b) Some observations on glass-knife making. *Stain Technol.* 52:305.

Weibull, C. (1970) Estimation of the thickness of thin sections prepared for electron microscopy. *Phillips Bull. EM* 45.

Weiner, S. (1959) A new method of glass knife preparation for thin section microtomy. *J. Biophys. Biochem. Cytol.* 5:175.

Williams, M. A., and G. A. Meek. (1966) Studies on thickness variation in ultrathin sections for electron microscopy. *J. Roy. Micros. Soc.* 85:337.

Willis, R. A. (1971) In situ measurement of electron microscope section thickness by use of a high angle tilting stage. *J. Anat.* 109:345.

Wyatt, J. H. (1974) An illumination system for specimen block alignment with the LKB Ultrotome Ultramicrotome. *J. Micros.* 101:207.

Zelander, T., and R. Ekholm. (1960) Determination of the thickness of electron microscope sections. *J. Ultrastr. Res.* 4:413.

9. Post-Staining

The need to enhance the electron density of biological materials has been stressed throughout previous discussions. Salts of heavy metals such as osmium and uranium applied as en bloc fixatives/stains considerably increase tissue density, and even more significant electron contrast may be induced by post-staining thin sections. Because sections are very small when compared to a tissue cube, post-stains will quickly penetrate and react with the cell (Peters et al., 1971), and will not be displaced by solvent extraction. Thin sections are extremely stable, meaning that distortion artifacts, as discussed in Chapter 5, are minimized (Lewis et al., 1974; Hayat, 1975). On the other hand, the tissue chemistry has been so significantly modified during fixation and embedding that site-specific stains may not act as such; site-selective stains are thus usually applied en bloc. General electron contrast is enhanced by staining first with uranyl acetate followed by a lead solution (Watson, 1958a); this sequence is referred to as "double salt staining."

Clearly electron stains are analogous to chromatic stains for light microscopy. However, chromatic dyes are frequently applied, for example, to distinguish nuclei from cytoplasm by staining with hematoxylin and eosin, respectively. Electron post-stains do not necessarily impart different contrast levels to organelles, which are identified on the basis of their morphology. To some extent organelles may also be identified by their position in the cytoplasm; for example, diplosomes are frequently seen near the Golgi apparatus. Nonetheless, organelles are best identified on the basis of their morphology.

The mechanism of staining with uranyl acetate was discussed earlier, in Chapter 7. It was introduced as a post-stain by Watson (1958a); and when it is used in conjunction with en bloc staining, nucleic acids exhibit excellent contrast (Huxley and Zubay, 1961; Zobel and Beer, 1961, 1965; Ellis et al., 1969; Bernhard, 1968, 1969; Franke and Falke, 1970; Lombardi et al., 1971; Papsidero and Braselton, 1973) as do membranes (Shah, 1969; Rothstein, 1970; Silva et al., 1971; Pollard and Korn, 1973), glycogen (Vye and Fischman, 1970), and lipids (Mumaw and Munger, 1969). Other uranyl compounds have been used as stains, but none are so popular as uranyl acetate. For example, uranyl formate has been applied as a negative stain (Leberman, 1965; Brack, 1973). Little research has been conducted, however, on the positive stain effects of other uranyl compounds.

A thin section mounted on a grid is post-stained with uranyl acetate simply by floating the grid on a drop of stain (0.5–1% concentration, aqueous) for ~5 min. Locke and Krishnan (1971) used hot alcoholic uranyl acetate as a post-stain, but Bhatnagar and Leeson (1975) noted that glycogen is masked. The grid is rinsed in distilled water and then again stained with lead. Avery and Ellis (1978) observed that a precipitate may form within tissue fixed with uranyl acetate (visible in the TEM) and found that it could be dissolved in thin sections exposed to 15% oxalic acid in methanol (30 min. at 40°C); the tissue is post-stained with hot uranyl acetate and rinsed in 0.25–0.5% aqueous oxalic acid for 10–15 sec. The oxalic acid rinse must be rapid to avoid destaining. The double-salt staining technique is the standard method for post-staining (e.g., Watson, 1958a,b; Marinozzi and Gautier, 1962; Frasca and Parks, 1965; Stockert et al., 1970; Hayat, 1970, 1975; Mumaw et al., 1976). Because uranyl acetate is toxic, the user must avoid direct contact with it and dispose of it safely (Darley and Ezoe, 1976; Thurston, 1978).

Watson (1958b) introduced lead salts as post-stains; he used lead hydroxide, and in doing so redefined contrast. Previously, phosphotungstic acid had been used as the sole stain for aldehyde-fixed tissues (Hall et al., 1945), but double salt fixation introduced far more contrast. The preparation of lead hydroxide is included in Chapter 11; basically it consists of lead acetate in sodium hydroxide. A major problem with this stain, however, is that an electron-dense lead carbonate precipitate forms upon exposure to airborne CO_2. Consequently other lead hydroxide formulations were introduced (Lever, 1960; Karnovsky, 1961; Bjorkman and Hellstrom, 1965). Concurrently other lead compounds were introduced as general-purpose stains. Lead acetate will convert to lead hydroxide within a section and thus minimize contamination, but the level of contrast is less than that obtained with Watson's stain (Dalton and Zeigel, 1960; Kushida, 1966). Kushida's (1966) preparation uses ethanol as the solvent, and this stain may be applied en bloc or as a post-stain. A different approach to minimizing lead carbonate is to use alkaline solutions of chelated lead; Millonig (1961) and Reynolds (1963) introduced, respectively, lead tartrate and lead citrate as electron stains. Microscopists have been content with these formulations since that time, but renewed interest is being shown in en bloc staining with lead aspartate (Yoshiyama et al., 1980). This stain produces sufficient contrast without embrittlement so that post-staining is unnecessary.

Reynolds (1963) proposed that in alkaline solution divalent lead salts ionize and produce the effective stain molecules:

$$Pb(OH)_2PbX_2 \rightleftarrows [Pb(OH)_2)Pb]^{++} + 2X^-$$

These cations are most numerous at pH 12, meaning that staining with very alkaline solutions is intense and rapid. On the other hand, Lever (1960) and Millionig (1961) proposed that alkaline solutions promote the formation of plumbite ions, $Pb(OH)_3^-$ or $Pb(OH)_4^=$, as the stain molecules. The citrate component of this stain prevents excessive lead carbonate formation. When this stain has been preceded by uranyl acetate treatment, the author has been satisfied with the degree of contrast, although others (e.g., Hayat, 1970) have noted that the imparted contrast is less than that of lead hydroxide. Williams and Adrian (1977) beautifully increased double-salt staining contrast by exposing the grid, after post-staining, for 30 sec to iodine vapors (also see Stewart, 1973). The degree of enhancement was identical to that of lead hydroxide.

The method of lead citrate staining is as follows (Reynolds, 1963): Place a few pellets of sodium hydroxide in a plastic Petri dish, close the dish, and wait a few minutes. This creates an NaOH atmosphere that avoids the formation of lead carbonate during the stain process.

Place a drop of stain per grid in the Petri dish and float the grids for ~15 min. with the dish closed. Rinse the grid by dipping it in 0.02 N NaOH followed by rinsing with distilled water. Blot the grids dry on filter paper and examine in the microscope.

When numerous grids are to be stained, this Petri dish method is inefficient. Inexpensive holders have been designed that the user may prepare in the laboratory, involving, e.g., modification of grid holders for submersion in liquid stain (e.g., Fisher, 1972; Hiraoki, 1972; Kushida, 1972; Robertson and Roberts 1972; Springer, 1974; Mollenhauer, 1975; Barate, 1976; Godkin, 1977; Gorycki, 1978).

Stock solutions of any lead stains must be tightly stoppered to prevent decomposition. Whenever the stock becomes cloudy, discard it. If amorphous, electron-dense clumps are observed on section surfaces, either the stock solution is deteriorated, or an improper procedure was employed. Another source of contamination is the sodium hydroxide wash; if opaque, acicular crystals are seen on the section, the NaOH is the source. In the author's lab, both lead citrate and NaOH are prepared fresh every 4–6 weeks.

Other researchers have developed lead citrate stains that are stable for up to 1 year (Sato, 1967), or react very rapidly (1–5 min.) with tissue sections and thus avoid contamination during the actual staining period (Venable and Coggeshall, 1965). Sundstrom and Mornstad (1975) also have prepared a stable alkaline lead citrate stain, which contains glucose.

Lead stains are reactive only when tissues have been osmicated; tissues fixed in aldehyde alone show moderate contrast when post-stained with uranyl acetate (Hayat, 1970). Lead increases the contrast of cell membranes (Frasca and Parks, 1965; Mumaw et al., 1976; Stockert et al., 1970), cellulose (Cox and Juniper, 1973), proteins and enzymes (Filshie and Rogers, 1961; Daems and Persijn, 1963; Livingstone et al., 1969; Rosenthal et al., 1969; Leskes et al., 1971; Pfeiffer et al., 1974), glycogen (Saltman et al., 1962; Daems and Persijn, 1963; Elbers et al., 1965; Paluello and Rosati, 1968; DeBruijn, 1970; Vye, 1971; Bhagwat and Wong, 1972; Simionescu et al., 1972), and phospholipids (Elbers et al., 1965; Adamson and Bowden, 1970; Dermer, 1970; Finlay-Jones and Papadimitriou, 1972; Gil, 1972; Pattle et al., 1972). In short, excellent general staining of the tissue will result from osmication, tertiary fixation, and post-staining with uranyl acetate, and post-staining with lead.

SURVEY OF OTHER POST-STAINS

A variety of additional post-stains are available for selective (not general) staining of given tissue constituents. Both Lewis et al. (1974) and Hayat (1975) have written books on post-staining. The following discussion touches on some of the more common site-selective stains; the reader should consult the above textbooks for other stains.

Phosphotungstic acid (PTA) has been encountered in previous chapters as a negative stain for particulate specimens and as a chromatic stain; PTA may also be used as a post-stain for collagen and elastin in connective tissue (Huxley, 1959), as well as surfaces of cells (Dermer, 1973). Silverman and Glick (1969) stained striated muscle with PTA and obtained very good contrast, but cell membranes were either faintly stained in negative image or not visible at all. Whereas uranyl and lead solutions will at least partially penetrate thin sections, PTA reacts strictly with the surface of the section. Thus, grids are submerged, not floated, on a large bead of stain. Aqueous or ethanolic solutions of concentration 0.5–5% are used for PTA staining; the duration is ~20 min. Locke and Krishnan (1971) rapidly increased contrast by using hot, alcoholic PTA as a post-stain.

Potassium permanganate ($KMnO_4$) initially was used as a heavy metal fixative for electron microscopy (Luft, 1956), but extracts phenomenal amounts of unfixed material. It is a very potent oxidizer (recall that $KMnO_4$ is used for OsO_4 removal in chromatic staining), and sections are stained with ~1% aqueous $KMnO_4$ for ~30 min. (Lawn, 1960). Membranes in general show good contrast (e.g., Parsons, 1961; Sutton, 1968). Embedding media containing nadic methyl anhydride cannot be used with $KMnO_4$ because the two will react (Reedy, 1965).

Silver is also a valuable post-stain for nervous tissue and cell surface carbohydrates; because it is frequently used in light microscopy and has sufficient density for TEM, correlative microscopy takes on a new meaning (Ribi, 1976). On the other hand, osmicated tissues will be distorted if silver-stained, meaning that only glutaraldehyde should be used (Hayat, 1970); if sufficient tissue is present, the user should prepare singly and doubly fixed tissues, then compare the results. Literally hundreds of published reports are available on silver staining; the reader should consult Hayat (1975) and Lillie (1977) for the various selective methods, some of which are included in chapter 11.

REFERENCES

Adamson, I. Y. R., and D. H. Bowden. (1970) The surface complexes of the lung. A cytochemical partition of phospholipid surfactant and mucopolysaccharides. *Am. J. Pathol.* 61:359.

Avery, S. W., and E. A. Ellis. (1978) Methods for removing uranyl acetate precipitate from ultrathin sections. *Stain Technol.* 53:137.

Barate, R. S. (1976) Grid staining–a useful device. *Stain Technol.* 51:193.

Bernhard, W. (1968) Une methode de coloration regressive a l'usage de la microscopie electronique. *C. R. Acad. Sci.* 267:2170.

—— (1969) A new staining procedure for electron microscopical cytology. *J. Ultrastr. Res.* 27:250.

Bhagwat, A., and P. Wong. (1972) Effect of pH in direct OsO_4 fixation on glycogen staining as shown by electron microscopy. *Stain Technol.* 47:39.

Bhatnagar, R., and T. S. Leeson. (1975) Masking of pleomorphic glycogen sites by methanolic uranyl acetate. *Stain Technol.* 50:213.

Bjorkman, N., and B. Hellstrom. (1965) Lead-ammonium acetate: A staining medium for electron microscopy free of contamination by carbonate. *Stain Technol.* 40:169.

Brack, C. (1973) Use of uranyl formate staining for the electron microscope visualization of DNA-protein complexes. *Experientia* 29:768.

Cox, G., and B. Juniper. (1973) Electron microscopy of cellulose in entire tissue. *J. Micros.* 97:343.

Daems, W. T., and J. P. Persijn. (1963) Section staining with heavy metals of osmium-fixed and formal-fixed mouse liver tissue. *J. Roy. Micros. Soc.* 81:199.

Darley, J. J., and H. Ezoe. (1976) Potential hazards of uranium and its compounds in electron microscopy: A brief review. *J. Micros.* 106:85.

Dalton, A. J., and R. R. Zeigel. (1960) A simplified method of staining thin sections of biological material with lead hydroxide for electron microscopy. *J. Biophys. Biochem. Cytol.* 7:409.

DeBruijn, W. C. (1970) Glycogen its chemistry and morphological appearance in the electron microscope. *J. Ultrastr. Res.* 42:29.

Dermer, G. B. (1970) The fixation of pulmonary surfactant for electron microscopy. *J. Ultrastr. Res.* 31:229.

—— (1973) Specificity of phosphotungstic acid as a section stain to visualize surface coats of cells. *J. Ultrastr. Res.* 45:183.

Elbers, P. F., P. H. J. T. Ververgaert, and R. Demel. (1965) Tricomplex fixation of phospholipids. *J. Cell Biol.* 24:23.

Ellis, L. F., R. M. VonFrank, and W. J. Kleinschmidt. (1969) Uranyl acetate and rotary shadowing to increase the contrast of small aggregated virus particles. *Proc. 27th Ann. EMSA Meet.*, p. 427.

Filshie, B. K., and G. E. Rogers. (1961) The fine structure of α-keratin. *J. Mol. Biol.* 3:784.

Finlay-Jones, J. M., and J. M. Papadimitriou. (1972) Demonstration of pulmonary surfactant by tracheal injection of tricomplex salt mixture: Electron microscopy. *Stain Technol.* 47:59.

Fisher, D. G. (1972) A holder for simultaneous fluid processing or carbon coating of electron microscope grids in lots of 10 or more. *Stain Technol.* 47:235.

Franke, W. W., and H. Falk. (1970) Appearance of nuclear pore complexes after Bernhard's staining procedure. *Histochemie* 24:266.

Frasca, J. M., and V. R. Parks. (1965) A routine technique for double-staining ultrathin sections using uranyl and lead salts. *J. Cell Biol.* 25:157.

Gil, J. (1972) Effect of tricomplex fixation on lung tissue. *J. Ultrastr. Res.* 40:122.

Godkin, S. E. (1977) Improved staining boxes for fast, uniform staining of ultrathin sections on grids. *Stain Technol.* 52:265.

Gorycki, M. A. (1978) An efficient staining method for ultrathin sections. *Stain Technol.* 53:11.

Hall, C. E., M. A. Jakus, and F. O. Schmitt. (1945) The structure of certain muscle fibrils as revealed by the use of electron stains. *J. Appl. Phys.* 16:459.

Hayat, M. A. (1970) *Principles and Techniques of Electron Microscopy*, vol. 1. Van Nostrand Reinhold, New York.

—— (1975) *Positive Staining for Electron Microscopy.* Van Nostrand Reinhold, New York.

Hiraoki, J. I. (1972) A holder for mass treatment of grids, adapted especially to electron staining and autoradiography. *Stain Technol.* 47:297.

Huxley, H. E. (1959) Some aspects of staining of tissue for sectioning. *J. Roy. Micros. Soc.* 78:30.

—— and G. Zubay. (1961) Preferential staining of nucleic acid-containing structures for electron microscopy. *J. Biophys. Biochem. Cytol.* 11:273.

Karnovsky, M. J. (1961) Simple method for staining with lead at high pH in electron microscopy. *J. Biophys. Biochem. Cytol.* 11:729.

Kushida, H. (1966) Staining of thin sections with lead acetate. *J. Electron Micros.* 15:93.

—— (1972) New device for staining thin sections. *J. Electron Micros.* 21:201.

Lawn, A. M. (1960) The use of potassium permanganate as an electron-dense stain for sections of tissue embedded in epoxy resin. *J. Biophys. Biochem. Cytol.* 7:197.

Leberman, R. (1965) Use of uranyl formate as a negative stain. *J. Mol. Biol.* 13:605.

Leskes, A., P. Siekevitz, and G. E. Palade. (1971) Differentiation of endoplasmic reticulum in hepatocytes. I. Glucose-6-phosphatase distribution in situ. *J. Cell Biol.* 49:264.

Lever, J. D. (1960) A method of staining sectioned tissues with lead for electron microscopy. *Nature* 186:810.

Lewis, P. R., D. P. Knight, and M. A. Williams. (1974) Staining methods for thin sections. In: *Practical Methods in Electron Microscopy.* Glauert, A. M. (ed.). American Elsevier, New York.

Lillie, R. D. (1977) *H. J. Conn's Biological Stains*, 9th ed. Williams and Wilkins Co., Baltimore.

Livingstone, D. C., M. M. Coombs, L. M. Franks, V. Maggi, and P. B. Gahan. (1969) A lead phthalocyanin method for the demonstration of acid hydrolases in plant and animal tissues. *Histochemie* 18:48.

Lombardi, L., G. Prenna, L. Okolicsangi, and A. Gautier. (1971) Electron staining with uranyl acetate. Possible role of free amino groups. *J. Histochem. Cytochem.* 19:161.

Locke, M., and N. Krishnan. (1971) Hot alcoholic phosphotungstic acid and uranyl acetate as routine stains for thin and thick sections. *J. Cell Biol. 50:550.*

Luft, J. H. (1956) Permanganate–a new fixative for electron microscopy. *J. Biophys. Biochem. Cytol.* 2:799.

Marinozzi, V., and A. Gautier. (1962) Fixation et coloration: etude des affinites des composants nucleoproteiniques pour l'hydroxide de plomb et l'acetate d'uranyle. *J. Ultrastr. Res.* 7:436.

Millonig, G. (1961) A modified procedure for lead staining of thin sections. *J. Biophys. Biochem. Cytol.* 11:736.

Mollenhauer, H. H. (1975) Post staining sections for electron microscopy: An alternate procedure. *Stain Technol.* 50:292.

Mumaw, V. R., M. P. Goheen, and B. L. Munger. (1976) The use of lead and uranyl ions as a stain in scanning electron microscopy. *Proc. 34th Ann. EMSA Meet.*, p. 314.

—— and B. L. Munger. (1969) Uranyl acetate-oxalate, an en bloc stain as well as a fixative for lipids associated with mitochondria. *Anat. Rec.* 169:383.

Paluello, F. M., and G. Rosati. (1968) The influence of fixation and dehydration on the isolated glycogen. *J. Microscopie* 7:275.

Papsidero, L. D., and J. P. Braselton. (1973) Ultrastructural localization of ribonucleoprotein on mitotic chromosome of *Cypreus alternifolius. Cytobiologie* 8:118.

Parsons, D. F. (1961) A simple method for obtaining increased contrast in Araldite sections by using postfixation staining of tissues with potassium permanganate. *J. Biophys. Biochem. Cytol.* 11:492.

Pattle, R. E., C. Schock, and J. M. Creasey. (1972) Electron microscopy of the lung surfactant. *Experientia* 28:286.

Peters, A., P. L. Hinds, and J. E. Vaughn. (1971) Extent of stain penetration in sections prepared for electron microscopy. *J. Ultrastr. Res.* 36:37.

Pfeiffer, U., E. Poehlmann, and H. Witschel. (1974) Kinetics of the accumulation of lead phosphate in acid phosphatase staining. *Proc. 2nd Int. Symp. EM and Cytochem.*, p. 25.

Pollard, T. D., and E. D. Korn. (1973) Electron microscopic identification of actin associated with isolated amoeba plasma membranes. *J. Biol. Chem.* 248:448.

Reynolds, E. S. (1963) The use of lead citrate at high pH as an electron opaque stain in electron microscopy. *J. Cell Biol.* 17:208.

Reedy, M. K. (1965) Section staining for electron microscopy. *J. Cell Biol.* 26:309.

Ribi, W. A. (1976) A Golgi-electron microscope method for insect nervous tissue. *Stain Technol.* 51:13.

Robertson, W. M., and I. M. Roberts. (1972) A simple device for the bulk staining and storage of ultrathin sections on grids. *J. Micros.* 95:425.

Rosenthal, A. S., H. L. Moses, D. L. Beaver, and S. S. Schuffman. (1969) Lead ion and phosphatase histochemistry. I. Nonenzymatic hydrolysis of nucleoside phosphated by lead ion. *J. Histochem. Cytochem.* 14:698.

Rothstein, A. (1970) A reappraisal of the action of uranyl ion on cell membranes. In: *Effects of Metals on Cells, Subcellular Elements, and Macromolecules.* Maniloff, J., F. R. Coleman, and M. W. Miller (eds.). Charles C. Thomas, Springfield, Ill.

Saltman, P., P. Charley, and B. Sarkar. (1962) Chelation of metal ions by sugars. *Fed. Proc.* 21:307.

Sato, T. (1967) A modified method for lead staining. *J. Electron Micros.* 16:133.

Shah, D. O. (1969) Interaction of uranyl ions with phospholipid and cholesterol monolayer. *J. Colloid Interf. Sci.* 29:210.

Silva, M. T., J. V. C. Melo, and F. C. Guerra. (1971) Uranyl salts as fixatives for electron microscopy. Study of the membrane ultrastructure and phospholipid loss in bacilli. *Biochem. Biophys. Acta* 233:513.

Silverman, C., and D. Glick. (1969) The reactivity and staining of tissue proteins with phosphotungstic acid. *J. Cell Biol.* 40:761.

Simionescu, N., M. Simionescu, and G. E. Palade. (1972) Permeability of intestinal capillaries; pathways followed by dextran glycogens. *J. Cell Biol.* 53:365.

Springer, M. (1974) A simple holder for efficient mass staining of thin sections for electron microscopy. *Stain Technol.* 49:43.

Stewart, M. (1973) Organic stains for electron microscopy. *J. Micros.* 97:381.

Stockert, J. C., O. D. Coleman, and P. Esponda. (1970) Ultrastructural morphology of teleophase prenucleolar bodies with the uranyl–EDTA–lead staining techniques. *J. Microscopie* 9:823.

Sundstrom, B., and H. Mornstad. (1975) Lead citrate-containing media for use at alkaline pH: Their stabilization with glucose and increased buffer strength. *Stain Technol.* 50:287.

Sutton, J. S. (1968) Potassium permanganate staining of ultrathin sections for electron microscopy. *J. Ultrastr. Res.* 21:424.

Thurston, E. L. (1978) Health and safety hazards in the SEM laboratory: Update 1978. *SEM, Inc.* 2:849.

Venable, J. H., and R. Coggeshall. (1965) A simplified lead citrate stain for use in electron microscopy. *J. Cell Biol.* 25:407.

Vye, M. V. (1971) A comparative evaluation of four methods for staining of glycogen in thin sections. *Lab. Invest.* 24:452.

—— and D. A. Fischman. (1970) The morphological alteration of particulate glycogen by en bloc staining with uranyl acetate. *J. Ultrastr. Res.* 33:278.

Watson, M. L. (1958a) Staining of tissue sections for electron microscopy with heavy metals. *J. Biophys. Biochem. Cytol.* 4:475.

—— (1958b) Staining of tissue sections for electron microscopy with heavy metals. II. Application of solutions containing lead and barium. *J. Biophys. Biochem. Cytol.* 4:727.

Williams, M. G., and E. A. Adrian. (1977) The use of elemental iodine to enhance staining of thin sections to be viewed in the electron microscope. *Stain Technol.* 52:269.

Yoshiyama, J. M., D. Goff, and J. Walton. (1980) Variations in sample preparation that affect contrast enhancement by lead aspartate. *Proc. 38th Ann. EMSA Meet.*, p. 654.

Zobel, C. R., and M. Beer. (1961) Electron stains. I. Chemical studies on the interaction of DNA with uranyl salts. *J. Biophys. Biochem. Cytol.* 10:335.

—— and M. Beer. (1965) The use of heavy metal salts and electron stains. *Int. Rev. Cytol.* 18:363.

10. Correlative Microscopy

INTRODUCTION

A variety of situations exist in which it is desirable to study a tissue specimen by light microscopy prior to electron microscopy. It is very simple to prepare and immediately examine a thick section during the fine trimming of the specimen block for ultramicrotomy: examination of the thick section will reveal the orientation and general morphology of the sample. For example, kidney glomeruli are readily distinguished from tubules, and in situations where the glomerulus is of interest but none are present in the thick section, simply cut deeper into the tissue and prepare slides until glomeruli are visible. Thin sections are then cut and examined. This saves a great deal of time that would otherwise be occupied by thin sectioning and TEM observation of nonpertinent materials.

The advantages of correlative light and electron microscopy have been very well demonstrated in pathology (DeBruijn and McGee-Russell, 1966; Lynn, 1975; Trump and Jones, 1978). Continuing the example of kidney, pathological differences that may be suspected at the light immunofluorescent level may be confirmed at the ultrastructural level (Spargo, 1975). The electron histology of tumors is producing volumes of material (e.g., Gyorkey et al., 1975). Until the mid-1960s diagnostic hematology was somewhat hazy, but combined TEM and SEM offered increased resolution and therefore more ready interpretation. This enhanced the potential of the light microscope because details previously unremarked upon (because of low resolution) were defined and therefore interpretable (Burns et al., 1975).

It will become clear that correlative microscopy is used in all areas of biology. Many of the techniques discussed below require only a few minutes' treatment, while more sophisticated methods are also available. The simpler methods involve direct application of a chromatic stain to a thick section, whereas, for example, site-selective staining may require resin and osmium removal.

Discussions of the light microscope as an instrument are available in Gray (1954), Barer (1968), Needham (1968), Bradbury (1976), Burrells (1977), and Rochow and Rochow (1978). Most college or university libraries have at least one of these texts available. A variety of chromatic staining methods are presented in Chapter 11, and more comprehensive reviews are available in Luna (1968), Lillie (1977), Clark (1973), and Hayat (1975). Hamm (1974) has written an excellent light histology test, and Rhodin (1974) correlates microscopy.

PREPARATION OF THICK SECTIONS

Earlier a quick method was presented for simultaneously fine trimming and thick sectioning specimens with an ultramicrotome (e.g., Reinius, 1966). The glass knives employed here need not be newly prepared; the level of resolution does not require freshly fractured edges. The block is mounted in the ultramicrotome as in the normal procedure for thin sectioning. A dry glass knife (i.e., one without a trough) is secured in the instrument, and the knife edge and specimen are aligned. Move the knife toward the specimen, and manually cycle the specimen: the knife is advanced 2–3 μm per stroke using the macrofeed advance. Trim away the excess embedding medium until the tissue is exposed. Trimming can be done rapidly, but decrease the cutting speed when the tissue is reached, and cut sections 2–3 μm thick. Use only good portions of the cutting edge; if scratches appear in the section, move to another area of the knife. Individual thick sections are picked up with watchmaker forceps at an edge, or a finely pointed object (e.g., needle) is slipped beneath the section, which is then transferred and floated on a drop of water on a microscope slide. Dry the slide over an alcohol flame for heat fixation. The section may be monochromatically stained with toluidine blue in sodium bicarbonate (Trump et al., 1961; Lynn 1965) or basic fuchsin (Winkelstein et al., 1963) to enhance general chromatic contrast as follows:

1. Place a few drops of stain over the tissue, and heat it gently for 30–60 sec.
2. Gently rinse the slide with distilled water, and dry it with heat.
3. Place on the slide a drop of the same formulation of embedding medium that the tissue is held in, coverslip, and polymerize at room temperature if a permanent mount is desired.
4. Examine the tissue with a light microscope.

Fairly thin sections tend to curl at the knife edge; these will flatten out during heat fixation. When nonpermanent mounts are preferred, e.g., for simple localization, use an immersion oil rather than embedding medium for coverslipping. Permanent mounts must be polymerized at room temperature to avoid leaching of the stain. The microscopist can observe the degree of stain uptake in step 1: if overstaining is observed (evaluate before coverslipping), the section may be destained simply by applying a drop of warm water to the slide, then rinsing with the same. All rinsing must be gentle to avoid dislodging the section. Alternatively, precoat the slide with a thin layer of Mayer's albumin adhesive or gelatin chrome alum adhesive (Jensen, 1962; Grimley, 1965); either of these treatments increases the adhesion between the section and slide. Very thick sections (4–10 μm) require that the slide be precoated with embedding medium for good adhesion (Burnett, 1975; Braak, 1977).

The advantages of this method are that it is simple and rapidly performed, the glass knives are not as critical as those used in thin sectioning, and finally back-cutting is concurrent with thick sectioning. The major disadvantages of dry sectioning are that sections thinner than ~2 μm are difficult to remove from the knife edge without damage, and the sections may be dislodged from the microscope slide if vigorously rinsed. Other methods are available for thick sectioning, but all are lengthy procedures and apply to more specific situations (e.g., site-selective reactions).

Thinner sections (0.5–1 μm thickness) are cut using the conventional flotation fluid-filled trough for wet-knife thick sectioning. The knife edge and block face are aligned, but the macroadvance is in smaller increments. Section thickness is evaluated by interference color, with green associated with ~1 μm. Place a drop of water on a microscope slide, and gently pick up the sections by sliding a thin glass rod (≤1 mm diameter) beneath them, then lifting them up, and refloating them on the slide. Proceed with staining as given above.

Thinner sections are proportionately more fragile, and it is easy to damage 0.5–1-μm sections during transfer to a slide. Gemmell and Henrikson (1970) developed a very large trough for direct mounting of individual sections or ribbons on a slide. The overall dimensions of the trough are 55 X 30 X 25 mm; a clean slide is submerged in the trough well, positioned beneath the sections, and lifted upward. The sections float on the slide and are then heat-fixed. The advantage of this technique is that relatively thin serial sections may be examined; on the other hand, some practice is necessary to transfer the complete ribbon to a slide.

CHROMATIC STAINING

The chromatic staining of tissues prepared for electron microscopy is not as well defined as conventional histological staining. Quantitative data on the influence of fixative and buffer composition on chromatic staining are lacking, but some qualitative factors have been defined. For example, osmium fixation may block potentially reactive sites for a chromatic stain; in some situations the thick section requires oxidation to remove reduced osmium (e.g., Munger, 1961; Shires et al., 1969). Likewise, polymerized resins must be impermeable for TEM, but this same characteristic can prevent penetration of the chromatic stain. Polymerized epoxy and polyester resins are insoluble in conventional organic solvents; thus the thick sections are pretreated by various means to render them soluble. Oxidation and polymer removal serve to increase reactivity and permeability to chromatic stains, respectively. Another parameter influencing chromatic staining is buffer type. Bryant and Watson (1967) compared the effect of hematoxylin and methylene blue on tissue fixed with phosphate- or veronal acetate–buffered osmium. They observed more intense staining with phosphate buffers. In these situations, it must be realized that optimal tissue preparation for TEM, not light microscopy, is the goal here. Always work toward optimal TEM criteria, then work around these parameters for chromatic staining. This is readily done by altering chromatic staining conditions (e.g., pH, temperature, and duration). If dissatisfied with these results, refer to, for example, Luna (1968) or Lillie (1977) for conventional light histological methods. Whenever sections are to be prepared with these methods, they must be secured on a slide by one of the adhesives discussed above.

A radically different but very desirable approach is to en bloc stain specimens with a material that functions as both an electron and a chromatic stain. Pourcho and Bernstein (1978) stained tissues with malachite green during primary glutaraldehyde fixation and obtained selective stain uptake by ribosomes, myofilaments, and lipids. The staining of nervous tissue by silver has long been used in both light microscopy and EM (e.g., Lillie, 1977); Ribi (1976) used silver for both microscopic examinations.

Osmium is extracted from thick sections by exposure to various oxidizing agents. Sevier and Munger (1968) used oxone, Shires et al. (1969) treated sections with potassium permanganate followed by oxalic acid, Pool (1969, 1973) employed hydrogen peroxide, and Heath (1970) used performic acid as the oxidizer. All of these methods are included in Chapter 11. The basic mechanism involved in oxidation is removal of osmium from basophilic structures, which are then free to react with chromatic stains (Hayat, 1975).

The desired degree of oxidation is at that point where all osmium is displaced; that is, the tissue section is not underoxidized (some osmium remains) or overoxidized (other chemical bonds are disrupted). The former situation decreases the intensity of staining, whereas the latter results in nonspecific or irregular staining (Sevier and Munger, 1968; Shires et al., 1969; Pool, 1969, 1973; Heath, 1970). Consequently, the optimal degree of oxidation must be determined for tissues by varying its duration. This procedure is required when polychromatic or site-specific staining is desired (e.g., Van Reempts and Borgers, 1975; Jha, 1976; Warmke and Lee, 1976).

Epoxy and polyester resins are permeable to basic and acid aqueous stains, respectively, and both resin types react nonselectively with alcohol-based stains. For example, epoxy resins may be stained with basic fuchsin or toluidine blue (Winkelstein et al., 1963; Chandra and Skelton, 1964; Goldstein, 1965; Lynn, 1965; Aoki and Gutierrez, 1967), but staining with hematoxylin and eosin requires polymer removal (Lane and Europa, 1965).

Shortly after the introduction of the epoxy resins, researchers ex-

Table 10-1. Chromatic staining of specimens embedded for TEM.

	STAIN	REACTIVE SITE	REFERENCE
A. ARALDITE	Azure II	General	Jeon, 1965
	Azure B, Alkaline fast green, Feulgen reaction	Flagellates	Dodge, 1964
	Basic fuchsin and crystal violet	Juxtaglomerular cells	Lee and Hooper, 1965
	Methylene blue–basic fuchsin	Peripheral nerve	Aparicio and Marsden, 1969
	Potassium permanganate	General	Grimley, 1965
	Del Rio Hortega's silver method	Golgi, membranes	Goldblatt and Trump, 1965
	Toluidine blue–basic fuchsin	Spermatozoa	Aoki and Gutierrez, 1967
	Tribasic staining	General	Grimley, 1964
	General	General	Richardson et al., 1960; Grimley et al., 1965
	General	Nervous tissue	Berkowitz et al., 1968
B. EPON	Aniline blue	Collagen, mucopolysaccharides	Biagini et al., 1972
	Azure B, Alkaline fast green, Feulgen reaction	Flagellates	Dodge, 1964
	Azure II	General	Jeon, 1965
	Basic fuchsin	General	Winkelstein et al., 1963
	Basic fuchsin–crystal violet	Juxtaglomerular cells	Lee and Hopper, 1965
	Basic fuchsin–methylene blue	General	Huber et al., 1968
	Hematoxylin and eosin	Pancreatic islets	Munger, 1961
	Hematoxylin and safranin	General	Schartz and Schecter, 1965
	Hematoxylin and eosin, PAS, PTA H	General	Lane and Europa, 1965
	Methylene blue–azure II–basic fuchsin	General	Humphrey and Pittman, 1974

Table 10-1. Continued

	STAIN	REACTIVE SITE	REFERENCE
	Methylene blue–basic fuchsin	Peripheral nerve	Aparicio and Marsden, 1969
	Paragon method	General	Martin et al., 1966; Spurlock et al., 1966; Lynn et al., 1966
	Paraphenylenediamine	General	Estable-Puig et al., 1965
	PAS/toluidine blue or PAS/silver methenamine	General	Cardno and Steiner, 1965
	Silver	Golgi, membranes	Movat, 1961; Ribi, 1976
	Del Rio Hortega's silver method		Goldblatt and Trump, 1965
	Sudan blue and Nile blue	General	McGee-Russell and Snale, 1963
	Toluidine blue	General	Trump et al., 1961; Lynn, 1965; Chandra and Skelton, 1964; Reinius, 1966
	Toludine blue–basic fuchsin	Spermatozoa	Aoki and Gutierrez, 1967
C. VESTOPAL W	General	General	Gautier, 1960; Moe et al., 1962; Schwalbach et al., 1963
	Giemsa	General	Thoenes, 1960
	PAS	Aldehyde rx sites	Tzitsikas et al., 1961, 1962
D. MARAGLAS	Paragon method	General	Spurlock et al., 1966
	Toluidine blue	General	Spurlock et al., 1966; Bennett and Radminska, 1966
E. GMA	Acid fuchsin, Toluidine blue	General	Ashley and Feder, 1966
	Methylene blue, basic fuchsin	General	Aparicio and Marsden, 1969
	General	Plants	Feder and O'Brien, 1968

pressed the fear that polymer removal would disrupt the tissue (Trump et al., 1961) or adversely affect selective staining (Munger, 1961). Subsequently, atypical solvents have been developed that will not distort the tissue, and very intense chromatic staining is possible.

The high degree of cross-linking present in polymerized epoxy resins prevents their ready solubilization. However, reaction with the polymer ester linkages by hydrolysis with potassium hydroxide in a solution of methanol and benzene (Fisch et al., 1956) or hydrolysis with sodium methoxide (Mayor et al., 1961) renders the medium soluble. The preparation of these reagents is dangerous, however, so Lane and Europa (1965) developed sodium hydroxide in ethanol, and Imai et al. (1968) and Snodgress et al. (1972) determined that potassium hydroxide in a mixture of methanol, acetone, and benzene works well. The most rapid method is halogenation with bromine vapors (Yensen, 1968), which renders the polymer soluble in ethanol; extreme care must be observed in handling bromine because it is extremely caustic. These methods are included in Chapter 11; see Winborn (1976) for more information. Baumann and Mendell (1974) have noted that Spurr is dissolved in propylene oxide during ultrasonic treatment, but the sections may be dislodged.

Following oxidation and/or resin removal, the sections should be stained immediately to avoid contamination. Examples of staining methods are presented in Table 10-1; general references for staining animal tissues are Grimley (1964), Grimley et al. (1965), Cardno and Steiner (1965), Huber et al. (1968), Aparicio and Marsden (1969); for plants, see O'Brien (1965), O'Brien and Thimann (1967), Feder and O'Brien (1968), and Kosaki (1973).

REFERENCES

Aoki, I. A., and L. S. Gutierrez. (1967) A simple toluidine blue–basic fuchsin stain for spermatozoa in epoxy sections. *Stain Technol.* 42:307.

Aparicio, S. R., and P. Marsden. (1969) A rapid methylene blue–basic fuchsin stain for semi-thin sections of peripheral nerve and other tissues. *J. Micros.* 89:139.

Ashley, C. A., and N. Feder. (1966) Glycol methacrylate in histopathology. *Arch. Path.* 81:391.

Barer, R. (1968) *Lecture Notes on the Use of the Microscope*. Blackwell Scientific Pub., Oxford.

Baumann, D., and J. R. Mendell. (1974) Method of re-embedding tissue for electron microscopy. *Stain Technol.* 49:118.

Bennett, D., and O. Radminska. (1966) Flotation-fluid staining: Toluidine blue applied to Maraglas section. *Stain Technol.* 41:349.

Berkowitz, L. R., O. Fiorello, L. Kruger, and D. S. Maxwell. (1968) Selective staining of nervous tissue for light microscopy following preparation for electron microscopy. *J. Histochem. Cytochen.* 16:808.

Biagini, G., P. Borsetti, and R. Laschi. (1972) Two rapid stainings for mucopolysaccharides and collagen in semi-thin sections. *J. Submicros Cytol.* 4:283.

Braak, E. (1977) Method of firmly attaching 4–10 μm thick Araldite serial sections to glass slides for light microscopic staining procedures. *Stain Technol.* 52:54.

Bradbury, S. (1976) *The Optical Microscope in Biology.* Camelot Press, Ltd., Southampton.

Bryant, V., and J. H. L. Watson. (1967) A comparison of light microscopy staining methods applied to a polyester and three epoxy resins. *Henry Ford Hospital Medical Bulletin* 15:65.

Burnett, B. R. (1975) A new method for serially mounting resin sections (Spurr) for light microscopy. *Stain Technol.* 50:288.

Burns, W. A., H. J. Zimmermann, J. Hammond, A. Howatson, A. Katz, and J. White. (1975) The clinician's view of diagnostic electron microscopy. *Human Pathol.* 6(4):467.

Burrells, W. (1977) *Microscopy Technique.* Fountain Press, London.

Cardno, S. S., and J. W. Steiner. (1965) Improvement of staining techniques for thin sections of epoxy-embedded tissue. *Am. J. Clin. Pathol.* 43:1.

Chandra, S., and F. R. Skelton. (1964) Staining juxtaglomerular cell granules with toluidine blue or with basic fuchsin for light microscopy after Epon embedding. *Stain Technol.* 39:107.

Clark, G. (1973) *Staining Procedures*, 3rd ed. Williams and Wilkins Co., Baltimore.

DeBruijn, N. C., and S. M. McGee-Russell. (1966) Bridging a gap in pathology and histology. *J. Roy. Micros. Soc.* 85:77.

Dodge, J. D. (1964) Cytochemical staining of sections from plastic-embedded flagellates. *Stain Technol.* 39:381.

Estable-Puig, J. F., W. C. Bauer, and J. M. Blumberg. (1965) Paraphenylenediamine staining of osmium-fixed plastic-embedded tissue for light and phase microscopy. *J. Neuropath. Exp. Neurol.* 24:531.

Feder, N., and T. P. O'Brien. (1968) Plant microtechnique: Some principles and new methods. *Am. J. Bot.* 55:123.

Fisch, W., W. Hofmann, and J. Koskikallio. (1956) The curing mechanism of epoxy resins. *J. Appl. Chem.* 6:429.

Gautier, A. (1960) Technique de coloration histologique de tissu inclus dans des polyesters. *Experientia* 16:124.

Gemmell, R. T., and R. C. Henrikson. (1970) A large trough for mounting thin sections on slides. *Stain Technol.* 45:135.

Goldblatt, P. J., and B. F. Trump. (1965) The application of Del Rio Hortega's silver method to Epon embedded tissue. *Stain Technol.* 40:105.

Goldstein, D. J. (1965) Relation of effective thickness and refractive index to permeability of tissue components in fixed sections. *J. Roy. Micros. Soc.* 84:43.

Gray, P. (1954) *The Microtomist's Formulary and Guide.* Blakiston, New York.

Grimley, P. M. (1964) A tribasic stain for thin sections of plastic-embedded, OsO_4-fixed tissues. *Stain Technol.* 39:229.

—— (1965) Selection for electron microscopy of specific areas in large epoxy tissue sections. *Stain Technol.* 40:259.

—— , J. Albrecht, and H. J. Mitchelitch. (1965) Preparation of large epoxy sections for light microscopy as an adjunct to fine structure studies. *Stain Technol.* 40:357.

Gyorkey, F., K. W. Min, I. Krisko, and P. Gyorkey. (1975) The usefulness of electron microscopy in the diagnosis of human tumors. *Human Pathol.* 6(4):421.

Hamm, A. W. (1974) *Histology*, 7th ed. J. B. Lippincott Co., Philadelphia.

Hayat, M. A. (1975) *Positive Staining for Electron Microscopy.* Van Nostrand Reinhold, New York.

Heath, E. (1970) The use of performic acid oxidation to facilitate differential staining of epoxy-embedded adinohypophipin. *Z. Zellforsch.* 107:1.

Huber, J. D., F. Parker, and G. F. Odland. (1968) A basic fuchsin and alkalinized methylene blue rapid stain for epoxy-embedded tissue. *Stain Technol.* 43:83.

Humphrey, C. D., and F. E. Pittman. (1974) A simple methylene blue–azure II-basic fuchsin stain for epoxy-embedded tissue sections. *Stain Technol.* 49:9.

Imai, J., A. Sue, and A. Yamaguchi. (1968) A removing method of the resin from epoxy-embedded sections for light microscopy. *J. Electron Micros.* 17:84.

Jensen, W. A. (1962) *Botanical Histochemistry.* W. H. Freeman and Co., San Francisco.

Jeon, K. W. (1965) Simple method for staining and preserving epoxy resin embedded animal tissue for light microscopy. *Life Sci.* 4:1839.

Jha, R. A. (1976) An improved polychrome staining method for thick epoxy sections. *Stain Technol.* 51:159.

Kosaki, H. (1973) Epoxy embedding, sectioning, and staining of plant material for light microscopy. *Stain Technol.* 48:111.

Lane, B. P., and D. L. Europa. (1965) Differential staining of ultrathin sections of Epon-embedded tissues for electron microscopy. *J. Histochem. Cytochem.* 13:579.

Lee, J. C., and H. Hopper, Jr. (1965) Basic fuchsin–crystal violet: A rapid staining sequence for juxtaglomerular cells embedded in epoxy resin. *Stain Technol.* 40:7.

Lillie, R. D. (1977) *H. J. Conn's Biological Stains*, 9th ed. Williams and Wilkins Co., Baltimore.

Luna, L. G. (ed.). (1968) *Manual of Histological Staining Methods of the Armed Forces Institute of Pathology*, 3rd ed. Blakiston, New York.

Lynn, J. A. (1965) Rapid toluidine blue staining of Epon-embedded and mounted "adjacent" sections. *Am. J. Clin. Pathol.* 44:57.

—— (1975) "Adjacent" sections–a bridge in the gap between light and electron microscopy. *Human Pathol.* 6:400.

—— , J. H. Martin, and G. J. Race. (1966) Recent improvement of histologic technics for the combined light and electron microscopic examination of surgical specimens. *Am. J. Clin. Pathol.* 45:704.

Martin, J. H., J. A. Lynn, and W. M. Nickey. (1966) A rapid polychrome stain for epoxy-embedded tissue. *Am. J. Clin. Pathol.* 46:250.

Mayor, H. D., J. C. Hampton, and B. Rosario. (1961) A simple method for removing the resin from epoxy-embedded tissue. *J. Biophys. Biochem. Cytol.* 9:909.

McGee-Russell, S. M., and N. B. Smale. (1963) On coloring Epon-embedded tissue sections with Sudan black B or Nile blue A for light microscopy. *Quart. J. Micros. Sci.* 104:109.

Moe, H., O. Behnke, and J. Rostgaard. (1962) Staining of osmium-fixed Vestopal-embedded tissue sections for light microscopy. *Acta Anat.* 48:142.

Movat, H. Z. (1961) Silver impregnation methods for electron microscopy *Am. J. Clin. Pathol.* 35:528.

Munger, B. L. (1961) Staining methods applicable to sections of osmium-fixed tissue for light microscopy. *J. Biophys. Biochem. Cytol.* 11:502.

Needham, G. H. (1968) *The Microscope, a Practical Guide*. Charles C. Thomas, Springfield, Ill.

O'Brien, T. P. (1965) Note on an unusual structure in the outer epidermal wall of the Avena coleoptile. *Protoplasma (Wien)* 60:136.

—— , and K. V. Thimann. (1967) Observations on the fine structure of the oat coleoptile, III. Correlated light and electron microscopy of the vascular tissue. *Protoplasma* (*Wien*) 63:443.

Pool, C. R. (1969) Hematoxylin-eosin staining of OsO_4-fixed Epon-embedded tissue: prestaining oxidation by acidified H_2O_2. *Stain Technol.* 44:75.

—— (1973) Prestaining oxidation by acidified H_2O_2 for revealing Schiff-positive sites in Epon-embedded sections. *Stain Technol.* 48:123.

Pourcho, R. B., and M. H. Bernstein. (1978) Malachite green: Applications to electron microscopy. *Stain Technol.* 53:29.

Reinius, S. (1966) Sectioning tissue for light microscopy with the Ultratome ultramicrotome. *Sci. Tools* 13:10.

Rhodin, J. A. G. (1974) *Histology: A Text and Atlas*. Oxford University Press, London.

Ribi, W. A. (1976) A Golgi-electron microscope method for insect nervous tissue. *Stain Technol.* 51:13.

Richardson, K. C., L. Jarrett, and E. H. Finke. (1960) Embedding in epoxy resins for ultrathin sectioning in electron microscopy. *Stain Technol.* 35:313.

Rochow, T. G., and E. G. Rochow. (1978) *An Introduction to Microscopy by Means of Light, Electrons, X-rays or Ultrasound.* Plenum Press, New York.

Schartz, A., and A. Schecter. (1965) Iron-hematoxylin and safranin O as a polychrome stain for Epon sections. *Stain Technol.* 40:279.

Schwalbach, G., K. G. Lickfeld, and H. Hoffmeister. (1963) Differentiated staining of osmium tetroxide–fixed, Vestopal W–embedded tissue in thin sections. *Stain Technol.* 38:15.

Sevier, A. C., and B. L. Munger. (1968) The use of oxone to facilitate specific tissue stainability following osmium fixation. *Anat. Rec.* 162:43.

Shires, T. K., M. Johnson, and K. M. Richter. (1969) Hematoxylin staining of tissues embedded in epoxy resins. *Stain Technol.* 44:21.

Snodgress, A. B., C. H. Dorsey, G. W. H. Bailey, and L. G. Dickson. (1972) Conventional histopathological staining methods compatible with Epon-embedded, osmicated tissue. *Lab Invest.* 26:329.

Spargo, B. H. (1975) Practical use of electron microscopy for the diagnosis of glomerular disease. *Human Pathol.* 6(4):405.

Spurlock, B. O., M. S. Skinner, and V. C. Kattine. (1966) A simple rapid method for staining epoxy embedded specimens for light microscopy with polychromatic stain. *Am. J Clin. Pathol.* 46:252.

Thoenes, V. W. (1960) Giemza-Farbung an Geweben mach Einbettung in Polyester ("Vestopal") und Methakrylat. *Z. Mikros.* 64:406.

Trump, B. F., and J. B. Jones (1978) *Diagnostic Electron Microscopy*, vols. 1 and 2. John Wiley and Sons, New York.

E. A. Smuckler, and E. D. Benditt. (1961) A method for staining epoxy sections for light microscopy. *J. Ultrastr. Res.* 5:343.

Tzitsikas, H., E. J. Rdzok, and A. E. Vater. (1961) Staining procedures for ultrathin sections of tissues embedded in polyester resin. *Stain Technol.* 36:355.

——, E. J. Rdzok, and A. E. Vater. (1962) Silver staining of ultra thin sections of tissues embedded in polyester resin. *Stain Technol.* 37:293.

VanReempts, J., and M. Borgers. (1975) A simple polychrome stain for conventionally fixed Epon-embedded tissues. *Stain Technol.* 50:19.

Warmke, H. E., and S. L. J. Lee. (1976) Improved staining procedures for semithin epoxy sections of plant tissues. *Stain Technol.* 51:179.

Winborn, W. B. (1976) Removal of epoxy resins from specimens for scanning electron microscopy. In: *Principles and Techniques of Scanning Electron Microscopy* 5:21. Hayat, M. A. (ed.). Van Nostrand Reinhold, New York.

Winkelstein, J. M. Menefee, and A. Bell. (1963) Basic fuchsin as a stain for osmium-fixed, Epon-embedded tissue. *Stain Technol.* 38:202.

Yensen, J. (1968) Removal of epoxy resin from tissue sections following halogenation. *Stain Technol.* 43:344.

11. Laboratory Procedures

IXATION AND EMBEDDING

ixation and Embedding: General Method

IN SITU PREFIXATION, 15 min.
Expose desired tissue and flood with chilled 2.5–3% phosphate (0.1 M containing 2.5 mM $CaCl_2$)-buffered glutaraldehyde.

↓

PRIMARY FIXATION 0–4°C, 1–2 hr
2.5–3% phosphate-buffered glutaraldehyde.
Mince the tissue into cubes $\leqslant 0.5\ mm^3$.

↓

BUFFER WASH 0–4°C
Several changes of chilled phosphate buffer.
Plants ~3 hr, animals ~15–30 min.

↓

POST-FIXATION 0–4°C, 1–2 hr
2% phosphate buffered OsO_4.

↓

WATER WASH 0–4°C
Several changes of distilled water.
Plants ~1 hr, animals, 15–30 min.

↓

AQUEOUS EN BLOC STAINING ↓
e.g., 1% uranyl acetate 10 min. ↓

DEHYDRATION
Graded series, 25, 50, 75, 100, 100%, ~15 min. per step, 0–4°C

↓

Ethanol ↓
Transition with 50% EtOH in propylene oxide and 100% propylene oxide 15–30 min. per step ↓

INFILTRATION
Graded series infiltration with acetone or proplyene oxide and embedding monomer to 100% monomer, ~30 min. per step, room temp.

↓

POLYMERIZATION
Cure blocks by heat according to standard method for each resin

Acetone

INFILTRATION
Water-miscible embedding media. Graded series → pure medium.

↓

POLYMERIZATION
UV light or heat

Fixation and Embedding: Animal Tissues

1. Fix the specimen in an aldehyde (below). Mince the tissues into 0.5-mm cubes during fixation. 0–4°C, 1.5–2 hr.
   a. The most commonly used buffers are 0.1 M cacodylate or phosphate containing 2.5 mM $CaCl_2$, the latter recommended for routine use.
   b. Adjust tonicity as desired.
   c. Fixative concentrations (see "Buffers and Fixatives" for formulations):
      2.5–3% glutaraldehyde
      2% formaldehyde/2.5% glutaraldehyde
      ~ 2% acrolein: use a maximum of 1 hr fixation

2. Rinse the tissue in cold buffer, three changes, 10 min. each.
3. Post-fix at 0–4°C with 2% OsO_4 for 1.5–2 hr. A buffer compatible with that used for primary fixation should be employed.
4. Rinse in several changes of cold distilled water for a total of 15 min.
5. Stain the specimen with ~1.5% uranyl acetate in distilled water for 10–15 min., cold. If acetone is used for dehydration, use ~1.5% uranyl acetate in 10% acetone for 10–15 min.
6. Dehydrate with a graded series (25, 50, 75, 100, 100%) of ethanol for 10–15 min. each, cold. Bring to room temperature in the final 100% ethanol rinse.
7. Substitute the ethanol with 50% propylene oxide in ethanol followed by 100% propylene oxide for 15 min. each.
8. Infiltrate with a 1:1 mixture of propylene oxide and an epoxy resin for 30 min., followed by two 100% changes of embedding medium.
9. Polymerize according to standard method.

Fixation and Embedding: Plant Tissues

1. Fix small pieces of tissue ($\leqslant 0.5\ mm^3$) in 2–4% glutaraldehyde buffered with 0.1 M phosphate buffer (pH ~6.8) for 1.5–2 hr at 0–4°C.
   Note: Substitution of acrolein, a mixture of acrolein/glutaraldehyde, or formaldehyde/glutaraldehyde (see "Sorenson's Phosphate Buffer") as the fixative or constant agitation may be used for more rapid fixation. If the tissues float, apply a mild vacuum using an aspirator.
2. Wash in cold 0.1 M phosphate buffer, three changes, for 1 hr each at 0–4°C.
   Note: Continual agitation or frequently changing the buffer accelerates washing.
3. Post-fix in 2% osmium tetroxide buffered with 0.1 M phosphate for 1.5–2 hr at 0–4°C. Agitate frequently.
   Note: Fixation at room temperature is complete within 1–1.5 hr.
4. Wash in cold distilled water, several changes, for a total of 1 hr.
5. Dehydrate in a graded series of acetone (25, 50, 75, 85, 95, 100, 100%) for 10–15 min. each at each step. Agitate. If desired, 2% uranyl acetate in 10% acetone may precede dehydration.

6. Infiltrate and embed in low viscosity Spurr embedding medium. (See Spurr procedure, below, under "Epoxy Resins.")
   Note: The rapid cure schedule may be used to accelerate preparation.

Fixation and Embedding: Kellenberger's Method for Suspended Bacteria

See below, "Kellenberger's Fixative for Bacteria," for solution formulations.

1. In a centrifuge tube, place suspended cells growing in tryptone medium and add 1% Kellenberger's OsO_4 to a final concentration of 0.1%.
2. Mix; then immediately centrifuge at 1800 *g* for 5 min.
3. Resuspend the pellet in 2% agar in tryptone, and let is solidify: then cut it into small cubes (0.5–1 mm^3).
4. Fix the cubes in 1.0 ml Kellenberger's OsO_4 fixative containing 0.1 ml of tryptone medium, overnight at room temperature.
5. Decant the OsO_4 and replace it with the uranyl acetate washing fixative for 2 hr at room temperature.
6. Dehydrate and embed according to standard procedure.
   Note: This method may be modified for bacteria growing on a solid substrate. After rinsing with balanced salt solution, simply begin at step 4 and continue processing as recommended.

Ryter, A., and E. Kellenberger (1958) *Z. Naturf.* 13:597.

Fixation and Embedding: Suspended Bacteria, Method II

1. Prepare a solution of 6% glutaraldehyde in 0.18 M cacodylate buffer.
2. Add a volume of the fixative equal to the volume of the cells suspended in growth medium containing 0.01% $CaCl_2 \cdot H_2O$. Shake the mixture and fix for 1 hr at 0–4°C.
3. Centrifuge, decant the fixative, and resuspend with 0.09 M cacodylate buffer. Wash with three changes of buffer for a total of 2–4 hr at 0–4°C.
4. If desired, the cells may be suspended in agar, or centrifugation at each step is required.

5. Post-fix if desired, dehydrate, and embed the specimens according to standard method.

Bacteria in Culture: Encapsulation in Agar

This method is used when centrifuged pellets of cells disintegrate during fixation and embedding.

1. Pre-fix the suspended cells using one of the methods recommended earlier. If the cells can tolerate heating to 45°C, the living culture may be used.
2. If cells were pre-fixed, resuspend them in a small volume of fixative. Unfixed cells are centrifuged and resuspended in a small volume of growth medium.
3. Prepare 2% aqueous agar and heat it in a water bath to 45°C; also warm the concentrated cell suspension to 45°C.
4. With a warm pipette transfer a small drop of agar to the specimen and gently shake it to form a homogeneous suspension.
5. Immediately transfer the specimen onto a cooler microscope slide (e.g., room temperature) using a warm pipette or simply by pouring the suspension.
6. After the agar solidifies (~5 min.), cut it into cubes (~0.5 mm^3), and continue processing.

Ryter, A., and E. Kellenberger (1958) *Z. Naturf.* 13:597.

BUFFERS AND FIXATIVES

Cacodylate Buffers and Fixatives

A. STOCK SOLUTION: 0.4 M Cacodylate Buffer

$Na(CH_3)_2AsO_2 \cdot 3H_2O$	42.8 g
Distilled water to make	500 ml

If sodium cacodylate is in anhydrous form, use 32.0 g and make solution up to 500 ml.

B. WORKING SOLUTION: 0.2 M Cacodylate Buffer

Stock 0.4 M buffer	50 ml	
0.2 M HCl	5.4 ml → pH 7.4	Select the
	8.4 ml → pH 7.2	appropriate pH
	12.6 ml → pH 7.0	for specimen
	18.6 ml → pH 6.8	type.
Distilled water to make	100 ml	

C. CACODYLATE-BUFFERED GLUTARALDEHYDE (2.5%)

Working 0.2 M buffer	50 ml
25% Glutaraldehyde (aqueous)	10 ml
Distilled water to make	100 ml

The final concentration of fixative is 2.5% in 0.1 M buffer. If desired, adjust osmolarity by adding sucrose, glucose, or sodium chloride.

D. CACODYLATE-BUFFERED FORMALDEHYDE (3%)

Working 0.2 M buffer	50 ml
40% Paraformaldehyde (aqueous)	7.5 ml
Distilled water to make	100 ml

E. CACODYLATE-BUFFERED (0.1 M) FORMALDEHYDE (2%)/GLUTARALDEHYDE (2.5%)
Refer to "Formaldehyde: Preparation of Aqueous Stock Solution," below.

Working 0.2 M buffer	50 ml
10% Paraformaldehyde (aqueous)	20 ml
25% Glutaraldehyde (aqueous)	10 ml
Distilled water to make	100 ml

F. CACODYLATE (0.1 M)-BUFFERED ACROLEIN (10%)

Working 0.2 M buffer	50 ml
Acrolein	10 ml
Distilled water to make	100 ml

G. CACODYLATE-BUFFERED (0.1 M) ACROLEIN (1%)/GLUTARALDEHYDE (2.5%)

Working 0.2 M buffer	50 ml
Acrolein	1 ml
25% Glutaraldehyde (aqueous)	10 ml
Distilled water to make	100 ml

H. CACODYLATE-BUFFERED (0.1 M) OSMIUM TETROXIDE (2%)

Working 0.2 M buffer	4 ml
4% OsO_4 (aqueous)	2 ml

I. CACODYLATE-BUFFERED (0.1 M) GLUTARALDEHYDE (0.8%)/OSMIUM TETROXIDE (0.7%)

Prepare 0.2 M cacodylate buffer at pH 7.4.
Prepare 2.5% glutaraldehyde and 10% osmium tetroxide in 0.1 M cacodylate buffer.
Cool to 0–4°C in an ice bath and combine the above fixatives, 1:2 of glutaraldehyde: osmium tetroxide.

Notes:
1. Arsenic-containing solutions are toxic and must be carefully handled.
2. The pH of the working buffer will drop over time; readjust to desired pH with 0.2 M HCl.
3. Refrigerate all solutions in glass-stoppered vessels.

Sabatini, D. D., K. Bensch, and R. J. Barrnett (1963) *Anat. Rec.* 142:274.

Collidine Buffer and Fixatives

A. STOCK SOLUTION

s-Collidine (pure)	5.34 g
Distilled water to make	100 ml

B. WORKING SOLUTION

Stock buffer solution	50 ml
1.0 N HCl	9 ml (adjust volume for desired pH)
Distilled water to make	100 ml

C. COLLIDINE-BUFFERED OsO_4 (2%)

Working buffer solution	20 ml
4% OsO_4 (aqueous)	10 ml

D. COLLIDINE-BUFFERED (0.2 M) FORMALDEHYDE (10%)

1. Preparation of formaldehyde

Paraformaldehyde powder	10 g
Distilled water	70 ml

Heat the above to 70°C for 20 min. with stirring. Add ~6 drops 1 N NaOH, while stirring, until the solution clears.

2. Final solution

Formaldehyde	70 ml
s-Collidine (pure)	2.4 ml
1 N HCl	5.0 ml
Distilled water to make	100 ml

Notes: 1. The stock and working solutions are stable indefinitely at room temperature.
2. Collidine-buffered glutaraldehyde may be prepared.

Bennett, H. S., and J. H. Luft (1959) *J. Biophys. Biochem. Cytol.* 6:113.

Kellenberger's Fixative for Bacteria: Osmium Tetroxide and Uranyl Acetate

A. STOCK VERONAL-ACETATE BUFFER

Sodium veronal	2.94 g
Sodium acetate (hydrated)	1.94 g

Sodium chloride	3.40 g
Distilled water to make	100 ml

B. WORKING KELLENBERGER BUFFER

Stock veronal-acetate buffer	5.0 ml
Distilled water	13.0 ml
0.1 N HCl	7.0 ml
1.0 M $CaCl_2$	0.25 ml

The desired pH is 6.0. Prepare just prior to use to avoid micro-organism contamination.

C. KELLENBERGER'S OsO_4 FIXATIVE (1%)

Kellenberger buffer	8 ml
4% OsO_4 (aqueous)	2 ml

D. WASHING FIXATIVE: URANYL ACETATE (0.5%)

Kellenberger buffer	10 ml
Uranyl acetate	0.05 g

E. TRYPTONE MEDIUM

Bacto-Tryptone (Difco)	1.0 g
NaCl	0.5 g
Distilled water	100 ml

Ryter, A., and E. Kellenberger (1958) *Z. Naturf.* 13:597.

Sorenson's Sodium Phosphate Buffer: Formulation I (0.1 M)

A. STOCK SOLUTION: 0.2 M Phosphate Buffer

1. Preparation of 0.2 M sodium phosphate monobasic

$Na_2HPO_4 \cdot 7H_2O$	26.85 g
Distilled water to make	500 ml

2. Preparation of 0.2 M sodium phosphate dibasic

$NaH_2PO_4 \cdot H_2O$	13.80 g
Distilled water to make	500 ml

B. WORKING SOLUTION: 0.1 M Phosphate Buffer

The following volumes of sodium phosphate monobasic and dibasic are combined for the desired pH. The mixture is then made up to 100 ml with distilled water.

	0.2 M Sodium Phosphate	
pH	Monobasic	Dibasic
6.8	25.5 ml	24.5 ml
7.0	19.5 ml	30.5 ml
7.2	14.0 ml	36.0 ml
7.4	9.5 ml	40.5 ml

The osmolarity at pH 7.2 is 226 mOsM. Adding 0.18 M sucrose to 0.1 M working buffer increases osmolarity to 425 mOsM.

C. PHOSPHATE-BUFFERED GLUTARALDEHYDE (2%)

Working 0.1 M solution	100 ml
25% Glutaraldehyde (aqueous)	8 ml

D. PHOSPHATE-BUFFERED OSMIUM TETROXIDE (2%)

Working 0.1 M solution	4 ml
4% OsO_4 (aqueous)	2 ml

E. PHOSPHATE-BUFFERED FORMALDEHYDE (3%)

Working 0.1 M solution	50 ml
40% Paraformaldehyde (aqueous)	7.5 ml

F. PHOSPHATE-BUFFERED GLUTARALDEHYDE (2.5%)/FORMALDEHYDE (2%)

Working 0.1 M solution	50 ml
10% Paraformaldehyde (aqueous)	20 ml
25% Glutaraldehyde (aqueous)	10 ml

G. PHOSPHATE-BUFFERED ACROLEIN (10%)

Working 0.1 M Solution	50 ml
Acrolein	10 ml

H. PHOSPHATE-BUFFERED ACROLEIN (1%)/GLUTARALDEHYDE (2.5%)

Working 0.1 M Solution	50 ml
Acrolein	1 ml
25% Glutaraldehyde (aqueous)	10 ml

Notes: 1. Refrigerate all solutions. When cloudiness or microorganism growth is observed, discard them.
2. The same phosphate buffer may be used for preparing both aldehyde and osmium fixatives.

Gomori, G. (1946) *Proc. Soc. Exp. Biol. Med.* 62:33.

Sodium Phosphate Buffer: Formulation II (0.135 M)

A. STOCK SOLUTION: 0.135 M Phosphate Buffer

$NaH_2PO_4 \cdot H_2O$	1.50 g
$Na_2HPO_4 \cdot 7H_2O$	15.20 g
Distilled water to make	500 ml

This buffer has pH 7.35, and osmolarity 298 mOsM.

B. The preparation of buffered fixatives is identical to the dilutions in "Sorenson's Sodium Phosphate Buffer: Formulation I." The only difference is that buffer tonicity is 0.135 M in this formulation.

Maunsbach, A. B. (1966) *J. Ultrastr. Res.* 15:242. Also see Karlsson, U., and R. L. Schultz (1965) *J. Ultrastr. Res.* 12:160 for additional phosphate buffers.

Veronal-Acetate Buffer and Osmium Tetroxide

A. STOCK SOLUTION: Veronal-Acetate Buffer

Sodium veronal (sodium barbital)	0.59 g
Sodium acetate (crystalline)	0.35 g
Distilled water to make	20 ml

B. STOCK SOLUTION: Ringer's Solution

Sodium chloride	40.25 g
Potassium chloride	2.10 g
Calcium chloride	0.90 g
Distilled water to make	500 ml

C. WORKING SOLUTION

Stock buffer solution	20.0 ml
Ringer's solution	6.8 ml
Distilled water	50.0 ml
0.1 N HCl	22.0 ml (adjust volume for desired pH)

D. VERONAL-ACETATE BUFFERED OSMIUM TETROXIDE (2%)

Working solution	19.5 ml
4% OsO_4 (aqueous)	10.0 ml

Note: 1. The veronal-acetate stock solution rapidly decomposes and should be prepared just before use.
2. Sodium veronal is a barbiturate and should be handled carefully.

Palade, G. E. (1952) *J. Exp. Med.* 95:285.

Formaldehyde: Preparation of Aqueous Stock Solution (10%)

Paraformaldehyde powder	10 g
Double distilled water	100 ml

1. Dissolve the powder in water by heating to ~60°C (working in a fume hood).
2. While the solution is warm, add 1.0 N NaOH dropwise until the solution clears.
3. Cool the solution before use.
4. Refrigerate it in a glass-stoppered bottle. If a precipitate forms over time, discard it.

Note: This method ensures that the working solution of formaldehyde is methanol-free.

EPOXY RESINS

Araldite

A. EMBEDDING MEDIUM

Araldite 502	27 ml *or* 100 g
DDSA	23 ml *or* 75 g
DMP-30	1 ml *or* 2.3–3.5 g

The medium may be frozen without accelerator for a few months. The complete medium may be anhydrously frozen for ∿5 days without severely increasing the viscosity.

B. METHOD

1. Dehydrate the specimen with acetone or ethanol.
2. If ethanol is used, substitute with propylene oxide:

50% ethanol/propylene oxide	~15 min	
100% propylene oxide	~15 min	
50% propylene oxide/Araldite	~30 min	Agitate
100% Araldite, two changes	~30 min each	

3. If acetone is used, infiltrate with Araldite:

50% acetone/Araldite	~30 min	Agitate
100% Araldite, two changes	~30 min each	

4. Embed in oven-dried capsules with fresh Araldite
5. Polymerization schedules:
   a. Overnight at 35°C ⟶ full day at 45°C ⟶ full day at 60°C
   b. 60–70°C for 12–24 hr
   c. 75–80°C for 2–3 hr, overnight at 60°C

Glauert, A. M., G. E. Rogers, and R. H. Glauert (1956) *Nature* 178:803.

Araldite/Poly/Bed*

A. EMBEDDING MEDIUM–STANDARD MOLLENHAUER

Poly/Bed 812	10 ml
Araldite 502	6 ml
DDSA	18 ml
DMP-30	0.52 ml

B. METHOD: identical to that for Araldite.

Mollenhauer, H. H. (1964) *Stain Technol.* 39:111.

Poly/Bed 812 (copyright, Polysciences 1980)*

A. EMBEDDING MEDIUM

Poly/Bed 812	21 ml
DDSA	13 ml
NMA	11 ml
DMP-30	0.7 ml

Frozen storage without accelerator is advised. Warm to room temperature before use. The accelerator constitutes 1.5–2% of the volume.

B. METHOD: identical to that for Araldite.

Rapid Epon Embedding: Semihydrated Medium

A. EMBEDDING MEDIUM

Epon 812	6.0 ml
Distilled water	0.48 ml

*Also see other suppliers (e.g., Ernest Fullam Inc.) for similar formulation.

Epon 815	4.0 ml
DMP-30	0.9 ml

B. METHOD

1. Dehydrate specimens in a graded series of ethanol up to 90% ethanol.
2. Infiltrate with ethanol/propylene oxide up to 100% propylene oxide.
3. Infiltrate with a graded series of propylene oxide/embedding medium, up to pure embedding medium.
4. Polymerize at 55°C for 4 hr *or* at 60°C for 2 hr.

Shinagawa, Y., Y. Shinagawa, and S. Uchida (1980) *Proc. 38th Ann. EMSA Meet.*, p. 642.

Spurr Low-Viscosity Embedding Medium

A. EMBEDDING MEDIA

	Standard firm	Hard	Soft	Rapid cure
VCD	10.0 g	10.0 g	10.0 g	10.0 g
DER-736	6.0	4.0	7.0	6.0
NSA	26.0	26.0	26.0	26.0
DMAE	0.4	0.4	0.4	1.0
Polymerization at 70°C	8 hr	8 hr	8 hr	3 hr

Stock solutions may be anhydrously frozen for ~3 months.

B. METHOD

1. Dehydrate in acetone or ethanol.
2. Infiltrate with acetone/Spurr or ethanol/Spurr:
   a. Add a quantity of pure medium equal to the volume of 100% dehydration reagent in the vial. Swirl to homogenize liquid and agitate frequently for ~30 min.

b. Add another equal volume of Spurr, swirl, and agitate frequently for ~30 min.
3. Decant the mixture and infiltrate with 100% Spurr; agitate frequently for ~30 min. Repeat with fresh medium; agitate 30 min.
4. Embed in oven-dried capsules.
5. Polymerize at 70°C for 8 hr overnight (but see "Rapid Cure" in above table).

Note: Contact with this medium causes dermatitis.

Spurr, A. R. (1969) *J. Ultrastr. Res.* 26:31.

POLYESTER RESIN

Vestopal W

A. EMBEDDING MEDIUM

Vestopal W	100 ml
Benzoyl peroxide	1 ml
Cobalt naphthenate	0.5 ml

Very thoroughly combine the Vestopal W with the benzoyl peroxide (~30–60 min. stirring). Then add the cobalt naphthenate. (COMBINING BENZOYL PEROXIDE DIRECTLY WITH COBALT NAPHTHENATE YIELDS AN EXPLOSIVE MIXTURE!) The medium may be anhydrously refrigerated for a few months, but fresh solutions are recommended.

B. METHOD

1. Dehydrate in a graded series of acetone and water.
2. Infiltrate with a graded series of acetone and Vestopal:

30% Vestopal in acetone	30–60 min.
50% Vestopal in acetone	30–60 min.
80% Vestopal in acetone	30–60 min.
100% Vestopal	Several hours

3. Polymerize at 60°C for 12–24 hr.

Ryter, A., and E. Kellenberger (1958) *J. Ultrastr. Res.* 2:200.

WATER-MISCIBLE EMBEDDING MEDIA

Aquon

A. DERIVATION OF AQUON

1. Shake one volume of Epon 812 with two volumes of water.
2. Centrifuge the mixture until two layers have formed and the upper layer is clear.
3. Carefully remove the supernatant (Aquon precursor), and add anhydrous sodium sulfate until an excess remains undissolved after shaking. Let the mixture stand at room temperature until two layers form.
4. Remove the upper layer and centrifuge at 0°C until a clear supernatant separates.
5. Decant the supernatant (80% Aquon) from the excess sodium sulfate.
6. The Aquon is dehydrated by placing it in a shallow tray with magnesium perchlorate in a desiccator.
   a. The Aquon dries in 2–3 days with a final yield ~30% of the initial Epon 812 volume.
   b. The Aquon is stable if stored in a tightly stoppered container with a drying agent ($CaCl_2$ will react with the Aquon; use Linde molecular sieve, Grade 4A, 1/16" pellets).

B. STOCK EMBEDDING MEDIUM

Aquon	10 ml
DDSA	25 ml

This mixture is stable for 1 week at room temperature.

C. FINAL EMBEDDING MEDIUM

Stock embedding medium	10 ml
BDMA	0.1 ml

Prepare immediately prior to use.

D. METHOD

1. Following the post-fixation water wash at 4°C, infiltrate the specimen with a graded series of Aquon and water. Infiltration must be conducted at 4°C.

2. Infiltrate with pure embedding medium ~4 hr.
3. Embed the specimens.
4. Polymerize at 60°C for 4 days.
5. Store the polymerized blocks in a desiccator.

Gibbons, I. R. (1958) *Proc. 4th Int. Cong. EM (Berlin)* 2:55.

Durcupan

A. EMBEDDING MEDIUM

Durcupan	5 ml
DDSA	11.5 ml
Accelerator-960	1–1.2 ml
Dibutylphthalate	0.2–0.4 ml

Prepare fresh as needed.

B. METHOD

1. Following the post-fixation water wash, infiltrate as follows:

50% Durcupan in water	30 min.
70% Durcupan in water	30 min.
90% Durcupan in water	30 min.
100% Durcupan in water	60 min.
100% Durcupan in water	60 min.

2. Polymerize at 45°C for 24 hr.

Note: Contact with this medium causes severe contact dermatitis.

Staubli, W. (1960). *C. R. Seanc. Soc. Biol.* 250:1137.

Glycol Methacrylate (GMA)

A. EMBEDDING MEDIUM

97% GMA + 3% distilled water	7 parts
98% Butylmethacrylate + 2% Luperco	3 parts

Prepare by prepolymerization:

1. A small portion of the above mixture is heated while swirling until boiling just begins. Avoid water contamination.

2. While still swirling it, plunge it into an ice bath until the temperature of the mixture is ~2°C.
3. Repeat steps 1 and 2 until the GMA at 0–4°C has the viscosity of a thick syrup.

B. METHOD

1. Following the post-fixation water wash, infiltrate at 0–4°C as follows:

80% GMA in distilled water	20 min.
97% GMA in distilled water	20 min.
100% GMA, several changes	~12 hr

2. Embed the specimens in gelatin (not polyethylene) capsules. Air must be displaced; the GMA will not polymerize in its presence.
3. Polymerize with UV light (λ >3150 Å) for 24–48 hr at 1–3°C.
4. Thin sections are mounted on carbon- or plastic-supported grids.

Note: Contact with this medium causes severe dermatitis.

Leduc, E. H., and W. Bernhard (1967) *J. Ultrastr. Res.* 19:196.

Hydroxypropyl Methacrylate (HPMA)

A. EMBEDDING MEDIUM

HPMA	30 ml
Azobisobutyronitrile	0.3 ml

Preparation by prepolymerization is identical to that given above under "Glycol Methacrylate."

B. METHOD

1. Following the post-fixation water wash, infiltration at 0–4°C with constant agitation is as follows:

80% HPMA in water	1 hr
80% HPMA in water	1 hr
97% HPMA in water	1 hr
97% HPMA in water	1 hr
100% HPMA	1 hr

2. Polymerization is by one of two methods:

	UV light at 10°C	12–24 hr
or:		
	56°C	2–3 days

Leduc, E., and S. J. Holt (1965) *J. Cell Biol.* 26:137.

POST-STAINS

Lead Hydroxide I (Watson, 1958)

A. PREPARATION

Lead acetate, $Pb(CH_2H_3O_2)_2$	8.26 g
Distilled water	15 ml

1. Dissolve the lead acetate by vigorous shaking. Add 3.2 ml of 40% NaOH by blowing with a pipette. Agitate the mixture.
2. Centrifuge; discard the supernatant; resuspend the lead in ~18.2 ml distilled water.
3. Repeat step 2; the supernatant is the stain.
4. Tightly seal the stain solution; saturation is maintained with a small amount of sediment.

B. METHOD

1. Float grids on a drop of stain for ~15 min. in a closed container.
2. Rinse several times in distilled water.

Watson, M. L. (1958) *J. Biophys. Biochem. Cytol.* 4:475.

Lead Hydroxide II (Lever, 1960)

A. PREPARATION

Lead hydroxide, $Pb(OH)_2$	1 g
Distilled water	100 ml

1. Combine the lead with water and bring the solution to a boil.

2. Cool it to room temperature and filter.
3. Add 2.0 N NaOH dropwise until the solution clears. Agitate the solution.

B. METHOD

1. Float grids on a drop of stain for 5 min. in a closed container.
2. Wash in 1% KOH.
3. Wash in distilled water.

Lever, J. D. (1960) *Nature* 186:810.

Lead Citrate I (Reynolds, 1963): Recommended Lead Stain

A. PREPARATION

Lead nitrate, $Pb(NO_3)_2$	1.33 g
Sodium citrate, $Na_3(C_6H_5O_7) \cdot 2H_2O$	1.76 g
Distilled water	30.0 ml

1. Thoroughly agitate the above chemicals in a glass-stoppered flask for ~20 min.
2. Add 8.0 ml 1 N NaOH, dilute to 50 ml with distilled water, and mix by inversion. The solution will clear and is ready for use.

B. METHOD

1. Place several NaOH pellets in a plastic Petri dish. Replace the lid and wait ~1 min.
2. Place drops of stain (1 drop/grid) in the dish, and immediately cover the dish.
3. Quickly introduce grids and float them on the drops for 12–15 min.
4. Wash the grids in 0.02 N NaOH.
5. Wash the grids in distilled water.

Note: Aqueous uranyl acetate post-staining should precede step 1 of the method: float the grids on drops of ~1% aqueous uranyl acetate for ~5 min, rinse in distilled water, blot dry, and proceeed with lead staining.

Reynolds, E. S. (1963) *J. Cell Biol.* 17:208.

Lead Citrate II (Venable and Coggeshall, 1965)

A. PREPARATION

Lead citrate, $Pb_3(C_6H_5O_7)_2 \cdot 3H_2O$	0.4 g
Distilled water	100.0 ml
10 N NaOH	1.0 ml

Agitate the above chemicals in a glass-stoppered vial.

B. METHOD

1. Stain by flotation for 1–5 min.
2. Rinse with distilled water.

Venable, J. H., and R. Coggeshall (1965) *J. Cell Biol.* 25:407.

Lead Citrate III (Sato, 1967)

A. PREPARATION

Lead nitrate, $Pb(NO_3)_2$	1.5 g
Lead acetate, $Pb(C_2H_3O_2)_2$	1.5 g
Lead citrate, $Pb_3(C_6H_5O_7)_2 \cdot 3H_2O$	1.5 g
Distilled water	90 ml

1. Mix the above chemicals and heat to 40°C for 1 min. while stirring.
2. Add 3 g sodium citrate; stir for 1 min.
3. Add 24 ml sodium hydroxide and 20 ml distilled water.

B. METHOD

1. Immerse grids in stain (use above formulation or dilute 1:7) ~10 min.
2. Wash in distilled water.

Sato, T. (1967) *J. Electron Micros.* 16:133.

Lead Tartrate (Millonig, 1961)

A. PREPARATION

NaOH	12.5 g
Potassium-sodium-tartrate, $KNaC_4H_4O_6 \cdot 4H_2O$	5 g
Distilled water to make	50 ml

1. Mix the above chemicals to form the stock solution.
2. Dilute 0.1 ml of the stock with 20 ml distilled water.
3. Heat and add 0.2 g lead hydroxide.
4. Cool and filter the solution.

B. METHOD

Stain as in "Lead Citrate I," above, for 5–10 min.

Millonig, G. (1961) *J. Biophys. Biochem. Cytol.* 11:736.

Lead Acetate I (Kushida, 1966)

A. PREPARATION

1. Add lead acetate $[Pb(CH_2H_3O_2)_2]$ to absolute ethanol until an excess is obvious.
2. Filter. The filtrate is the stain.

B. METHOD

1. Immerse grids in stain ~15 min.
2. Rinse in absolute ethanol.
3. Blot dry on filter paper.

Kushida, H. (1966) *J. Electron Micros.* 15:93.

Lead Acetate II (Bjorkman and Hellstrom, 1965)

A. PREPARATION

1. Saturated lead acetate solution in water:

Lead acetate	39 g
Distilled water	100 ml

2. Stain solution:

Saturated lead acetate	100 ml
Ammonium acetate, $NH_4C_2H_3O_2$	18.5 g

B. METHOD

1. Immerse grids in stain 20–45 min.
2. Wash with distilled water.

Bjorkman, N., and B. Hellstrom (1965) *Stain Technol.* 40:169.

Phosphotungstic Acid Post-stain

1. Prepare a 0.5–5% solution of PTA in distilled water or ethanol.
2. Immerse the grids in the stain ~20 min, hot.
3. Rinse with distilled water if aqueous PTA is used, or in absolute ethanol if it is used as the solvent.

Locke, M., and N. Krishnan (1971) *J. Cell Biol.* 50:550.

Potassium Permanganate Post-stain

1. Prepare 0.5–1% aqueous $KMnO_4$.
2. Stain the grids by floating ~30 min.
3. Rinse in distilled water.

Lawn, A. M. (1960) *J. Biophys. Biochem. Cytol.* 7:197.

CHROMATIC STAINING

Aniline Blue for Collagen

A. PREPARATION

Aniline blue	1 g
Acetic acid	16 ml
Distilled water	100 ml

B. METHOD

1. Stain sections with a few drops of toluidine blue solution for 1–2 min. at 80°C.
2. Rinse thoroughly in distilled water and counterstain with eosin-erythrosin solution for 40–60 sec at 80°C.
3. Rinse thoroughly in distilled water and counterstain with aniline blue for 30–45 sec at 80°C.
4. Rinse in distilled water.
5. Air-dry and coverslip.

Biagini, G., P. Borsetti, and R. Laschi (1972) *J. Submicro. Cytol.* 4:283.

Azure B for Intact Epon Sections of Plant Tissues

A. PREPARATION

Azure B	0.6 g
Sodium bicarbonate	3 g
Distilled water	300 ml

B. METHOD

1. Cut 0.5–1.0-μm-thick sections.
2. Transfer the sections to a drop of 10% acetone on a clean glass slide and heat slightly to expand the section and evaporate the acetone.
3. Flood the section with 0.2% azure B solution in 1% sodium bicarbonate at pH 9.0, place the slide on a hot plate at 50°C for 2–5 min., and then rinse rapidly in tap water.
4. Air-dry and coverslip with Epon. Do not use heat for polymerization.

C. RESULTS

Blue: nucleoli and primary walls.
Light blue: secondary walls.
Gray: cytoplasm.
Blue-gray: nuclei.
Blue-green: chloroplasts.

Hoefert, L. L. (1968) *Stain Technol.* 43:145.

Basic Fuchsin/Malachite Green/Iron Chloride Hematoxylin

A. PREPARATIONS

1. Basic fuchsin

Ethanol (50%)	5 ml
Basic fuchsin	2 g
Distilled water	45 ml

2. Malachite green

Ethanol (30%)	100 ml
Azure B	0.4 g
Malachite green	1 g
Aniline	1 ml

This stain does not dissolve completely; refrigerate for 3 days and use supernatant. Do not filter.

3. Iron chloride hematoxylin

Distilled water,	75 ml
HCl (conc.)	0.5 ml
Ferric chloride crystals, ($FeCl_3 \cdot 6H_2O$)	0.62 g
Ferrous sulfate crystals, ($FeSO_4 \cdot 7H_2O$)	1.12 g

After the iron compounds have dissolved, add 25 ml freshly prepared 1% alcoholic solution of hematoxylin.

4. Phosphate buffer (0.2 M)

Solution A:

Sodium phosphate monobasic, ($Na_2HPO_4 \cdot H_2O$)	2.78 g
Distilled water	100 ml

Solution B:

Sodium phosphate dibasic, ($NaH_2PO_4 \cdot 7H_2O$)	5.37 g
Distilled water	100 ml

The desired pH can be obtained by mixing the two solutions in the ratios given below and diluting to a total volume of 2000 ml.

Solution A (ml)	Solution B (ml)	pH
39.0	61.0	7.0

B. METHOD

1. Cut 0.5–2.0-μm-thick sections and heat-fix them to a clean glass slide.
2. Immerse the slide in 2% NaOH in absolute ethanol for 10 min. to dissolve the plastic.
3. Wash the slide in four changes of absolute ethanol for 4 min. each, and then place it in phosphate buffer at pH 7 for 5 min.
4. Rinse the slide in three changes of distilled water and then place it in buffer at pH 4 for 5 min.
5. Wash for 5 min. in tap water and stain in iron hematoxylin for about 20 min. Staining should be controlled microscopically. The duration of staining is dependent upon section thickness.
6. Rinse rapidly in tap water and examine: chromatin and mitochondria must be light blue-gray and the cytoplasm clear; if overstained, destain with 4% ferric alum.
7. Stain in malachite green for 2 min.
8. Rinse rapidly in tap water.
9. Immerse in basic fuchsin on a hot plate at 37°C for 20–120 sec and rinse rapidly in tap water.
10. Air-dry and mount in Araldite. Press coverslip to obtain the thinnest possible layer of plastic and polymerize, preferably at room temperature.

C. RESULTS

Bright blue-green: myelin, lipid droplets, nucleoli, oligodendrocytes.
Purplish-pink: nuclei and astrocytes.

Berkowitz, L. R., O. Fiorello, L. Kruger, and D. S. Maxwell (1968) *J. Histochem. Cytochem.* 16:808.

Basic Fuchsin/Methylene Blue

A. PREPARATION

Sodium phosphate monobasic	0.5 g
Basic fuchsin	0.25 g
Methylene blue	0.2 g
Boric acid solution (0.5%)	15.0 ml
Distilled water	70.0 ml
NaOH (0.72%) (pH 6.8)	10.0 ml

The pH range of the solution should be 6–8. The solution is stable for several months. Shake it before using.

B. METHOD

1. Place about 1 ml of the staining solution on the glass slide with heat-fixed sections, and heat for 4–5 sec at 45–50°C.
2. Rinse in running tap water and air-dry.
3. If staining is insufficient, repeat steps 1 and 2.
4. Cover sections with a drop of immersion oil and seal with epoxy.

C. RESULTS

Red: mitochondria.
Pink: erythrocytes in blood vessels, glomeruli and tubules in kidney.
Brilliant pink: collagen.

Reddish-purple: elastic lamina, zymogen granules.
Bluish-purple: nuclei.

Sato, T., and M. Shamoto (1973) *Stain Technol.* 48:223.

Basic Fuchsin/Toluidine Blue O

A. PREPARATION

Polyethylene glycol	50 ml
0.2 N KOH	0.75 ml
Basic fuchsin	1.7 g
Toluidine blue O	0.3 g

Powder the stains before adding them to the solution.

B. METHOD

1. Place the glass slide with sections on a hot plate for 45 sec at 100°C.
2. While it is still on the hot plate, place a drop of staining solution on each section for 2–3 min.
3. Rinse the slide in 10% acetone.
4. Dry it over an alcohol lamp.

C. RESULTS

Deep blue to blue-purple: mitochondria, nucleus, plasma and nuclear membranes, muscle striations, cuticular structures, nerve sheaths, certain secretory bodies, dense bodies.
Deep violet: enzymatic secretions, some secretory vesicles.
Red: basement membrane, microvilli, nuclei, some secretory vesicles.
Grayish-green-violet: fat droplets.

Alsop, D. W. (1974) *Stain Technol.* 49:265.

Celestin Blue B—Cason's Solution for Connective Tissues

A. PREPARATION

1. Celestin blue B

Celestin blue B	0.3 g
5% iron alum	60 ml

a. Dissolve in distilled water, and boil the solution 2–3 min.
b. Cool the solution to room temperature, and filter it before use.
c. Add 7 ml glycerol.

2. Modified Cason solution

Orange G	2 g
Aniline blue (water-soluble)	3 g
Acid fuchsin	3 g

a. Consecutively dissolve the above in the given sequence in 200 ml of 0.1 N HCl.
b. Filter.

B. METHOD

1. Oxidize OsO_4-fixed, Epon-embedded sections in 5% H_2O_2 for 7 min.
2. Rinse in distilled water 1 min.
3. Stain in celestin blue B for 15 min. at 60°C.
4. Wash in running tap water for 5 min.
5. Stain in modified Cason solution 10 min. at room temperature.
6. Differentiate in 0.1 N HCl for 30 sec.
7. Rinse twice in distilled water.
8. Heat-dry.
9. Clear in xylene and mount.

C. RESULTS

Blue: elastic fibers.
Dark brown: nuclei.
Red: collagen fibers.
Yellow: cytoplasm.

VanReempts, J., and M. Borgers (1975) *Stain Technol.* 50:19.

Celestin Blue B-Eosin

A. SOLUTIONS

1. Iron alum (5%)
   Dissolve clear violet ferric ammonium sulfate in distilled water and store the solution in the cold.
2. Eosin (1%)

Eosin Y	10 g
Distilled water	1000 ml
Calcium chloride (anhydrous)	1 g
Thymic acid	a small crystal

   Dissolve eosin in water and bring solution to boiling in a 2000-ml beaker. Add calcium chloride while stirring and then lower heat to simmer for 1–2 min. Cool solution to room temperature and add thymic acid.
3. Celestin blue B

Celestin blue B	0.3 g
Solution A	60 ml
Glycerol	7 ml

   Dissolve celestin in Solution A and boil gently for 2–3 min. Cool to room temperature, filter, and add glycerol.

B. METHOD

1. Transfer 3-μm-thick sections to a glass slide and dry on a hot plate at 100°C for 10–15 min.
2. Remove the resin and reduced osmium according to standard procedures.
3. Stain sections with Solution C for 5–8 min.
4. Rinse thoroughly in distilled water.
5. Counterstain in Solution B for 15 min.
6. Dehydrate rapidly in ethanol.
7. Clear in equal parts of terpineol and xylene, followed by two changes of xylene.
8. Coverslip with resin.

C. RESULTS

Deep blue: nuclear chromatin.
Pink: cytoplasm.

Snodgress, A. B., C. H. Dorsey, G. W. H. Bailey, and L. G. Dickson (1972) *Lab. Invest.* 26:329.

Hematoxylin–Safranin for Plant Tissues

A. PREPARATION

1. 5% H_2O_2
2. 3% Ferric ammonium sulfate, acidified (mordant)

2-Methoxyethanol (methyl cellosolve)	50 ml
Distilled water	50 ml
Glacial acetic acid	1 ml
Sulfuric acid	0.12 ml
Ferric ammonium sulfate	3 g

3. Hematoxylin (Hellige CH-51 or Sigma 54C-0053)

2-Methoxyethanol	50 ml
Tap water	25 ml
Distilled water	25 ml
10% hematoxylin in absolute ethanol	10 ml

   Allow ∿2 weeks at room temperature for ripening.

4. Safranin O (Natl. NS-24 or Fisher CI-50240)

Safranin O	1 g
0.2 M Tris buffer, pH 9.0	100 ml

B. METHOD

1. Oxidize sections with 4% H_2O_2 for 10 min. at room temperature.
2. Rinse in running water.
3. Transfer to ferric ammonium sulfate in screw-capped Coplin jars and place in oven at 60°C for 30–60 min.
4. Rinse and transfer to hematoxylin solution in Coplin jars; oven-heat at 60°C 30–60 min.
5. Rinse and then stain in preheated (60°C) safranin O solution for 5 min.
6. Rinse in running tap water, then distilled water; heat-dry at 60°C.
7. Clear by flooding sections at room temperature with 1 : 1 clove oil and xylene for 30 sec.

8. Pass through three changes of pure xylene and mount with Permount.

C. RESULTS

Gray to black: basophilia.
Yellow: cell wall structures.

Warmke, H. E., and S. L. J. Lee (1976) *Stain Technol.* 51:179.

Methylene Blue/Azure II/Basic Fuchsin

A. PREPARATION

1. Methylene blue–Azure II

Methylene blue	0.13 g
Azure II or Azure A	0.02 g
Glycerol	10 ml
Methanol	10 ml
Phosphate buffer (pH 6.9)	30 ml
Distilled water	50 ml

2. Basic fuchsin

Basic fuchsin	0.1 g
Ethanol (50%) to make	10 ml

Prior to use, 3 ml of this solution should be diluted with 60 ml of distilled water.

3. Phosphate buffer

KH_2PO_4 (anhydrous)	9.078 g
$NaHPO_4$ (anhydrous)	11.876 g
Distilled water to make	1 liter

B. METHOD

1. Immerse slides with sections in Solution A (#1) at 65°C. Durations required are:

Maraglas with D.E.R.	0.5–5 min.
Maraglas with cardolite	2–5 min.
D.E.R. 332-732	3 min.
Epon–Araldite	5 min.

Spurr	20 min.
Araldite	30 min.
Epon	30–60 min.

2. Rinse in two changes of distilled water.
3. Transfer to Solution B (#2) for 0.5–5 min. (depending upon the embedding medium) at room temperature.
4. Rinse in three changes of distilled water.
5. Air-dry and coverslip with 1 drop of xylene and 1 drop of embedding medium.
6. Dry the mounting medium at room temperature; heat will fade the basic fuchsin.

C. RESULTS

Blue: nuclei.
Light blue: cytoplasm.
Pink to purple: goblet cell mucin.
Pink to red: glycogen and amylopectin.
Gray to light blue: lipid droplets.
Blue-green: erythrocytes.
Gray to pinkish-gray: brush border of intestinal epithelium.

Humphrey, C. D., and F. E. Pittman (1974) *Stain Technol.* 49:9.

Periodic Acid/Schiff/Toluidine Blue (PAS)

A. PREPARATION

1. Periodic acid: 1% aqueous, stored at 20°C.
2. Schiff reagent: commercially available, or: bring to a boil 200 ml distilled water. Add 1 g basic fuchsin and stir to dissolve. Cool to 50°C. Filter and add 20 ml normal hydrochloric acid. Cool to 25°C and add 1 g sodium bisulfite. Keep in the dark. The fluid may take 18–24 hr to become straw-colored; then it is ready for use. Store the solution cold in a tightly stoppered amber bottle. Discard it when a pink color appears. The solution will remain viable 2–3 weeks in use.
3. Sulfurous acid rinse

Sodium metabisulfite, 10% aqueous	6 ml

Normal HCl	5 ml
Distilled water	100 ml

Prepare just before use.

4. Toluidine blue O: 5% aqueous buffered with 5% sodium carbonate (pH 11). Mix well and heat until just boiling. Filter the solution through Whatman #5 filter paper while it is still hot. Cool it to 20°C before use. The solutions should be prepared fresh as needed.

B. METHOD

1. The sections are individually transferred (not on slides) to the surface of the periodic acid solution, contained in a low-form staining dish, and allowed to float on it for 15 min. at 20°C.
2. The sections are rapidly passed to the surface of two consecutive water rinses.
3. The Schiff reagent is brought to 20°C. The sections are floated onto its surface in a low-form staining dish, and allowed to stand at room temperature for 30 min. Increasing this time does not enhance staining.
4. The sections are transferred directly (without water rinsing) onto two consecutive sulfurous acid rinses for two min. each.
5. The sections are floated on water twice.
6. The sections are transferred onto the toluidine blue solution, and wet sections are picked up on slides and checked in a microscope at 1-min. intervals. When the desired intensity of staining has been reached (about 5 min.), the sections are floated twice on water.
7. The stained sections are placed in the center of the water surface in a Coplin jar. An uncoated, unheated slide is carefully inserted into the water to half of its length behind the floating sections. The slide is then withdrawn with a rapid upward and forward movement. Remove excess water with filter paper and air-dry.

Cardno, S. S., and J. W. Steiner (1965) *Am. J. Clin. Pathol.* 43:1.

Periodic Acid/Silver Methenamine (PASM)

A. PREPARATION

1. Periodic acid: 1% aqueous, stored at 20°C.
2. Silver methenamine:
   (a) 45 ml of 3% aqueous hexamethylenamine stored at 4°C.
   (b) 5 ml of 5% aqueous silver nitrate, stored at 20°C.
   (c) 5 ml of 0.05 M aqueous sodium borate solution, stored at 4°C.

   Bring Solutions (a) and (c) to room temperature. Add Solution (b) to Solution (a). The solution, initially milky, should clear upon mixing. Solution (c) is then added. The solution is mixed well, and filtered through Whatman #5 filter paper.
3. Sodium thiosulfate: 3% aqueous, stored at 20°C.

B. METHOD

1. The sections are transferred individually (not on slides) to the surface of the periodic acid solution, contained in a low-form staining dish, and allowed to float on it for 15 min. at 20°C.
2. The sections are rapidly passed to the surface of two consecutive water rinses.
3. The sections are transferred to the surface of the silver methenamine solution in a covered staining dish and placed in an incubator at 60°C for 20 min., and then transferred to an incubator at 45°C for 30 min. A section is then removed onto a water bath, picked up on a slide, and examined while still wet to determine the intensity of the stain. Similar periodic checks are made at 10–15-min. intervals. The examined sections can be further stained.
4. Individually float the sections on the sodium thiosulfate solution for 30 sec, and then onto two rinses of water.
5. Precleaned slides are coated with an extremely thin layer of Mayer's albumin fixative and heated to 60°C on a hot plate; they are used immediately upon removal from the hot plate.
6. Five to ten stained sections are placed in the center of the water surface in a Coplin jar. The heated slide is carefully inserted into the water to half of its length behind the floating

sections. The slide is then withdrawn with a rapid upward and forward movement. Remove excess water with filter paper.

7. Dry the slide on a hot plate at 60°C for about 15 min.; then mount with the embedding medium.

C. RESULTS

Black: carbohydrates.
Yellow to brown: all other structures.

Cardno, S. S., and J. W. Steiner (1965) *Am. J. Clin. Pathol.* 43:1.

Silver Methenamine for Basement Membranes of Kidney

A. SOLUTION

Silver methenamine

5% silver nitrate ($AgNO_3$)	5 ml
3% methenamine	100 ml

B. METHOD

1. Remove the epoxy resin according to standard procedures.
2. Place the slide in 0.5% periodic acid for 15 min., and wash it thoroughly in deionized water.
3. Prewarm the silver methenamine solution in a Coplin jar in hot water and the slide on a 94°C hot plate. Stain for 40–70 min. at 50–60°C. Control the degree of staining with a microscope.
4. Wash the slide in deionized water and put it into 0.05% gold chloride for 1–3 min.
5. Wash it with deionized water, and place it in 2.5% sodium thiosulfate for 1–3 min.
6. Wash with water, dry, and mount.

C. RESULTS

Black: glomerular basement membrane, Bowman membrane, basement membrane of tubules, internal elastica of arteries.
Black to dark brown: mesangial matrix and cytoplasmic matrix.
Dark brown: nuclear membrane.

Pale tan: nucleoplasm.
Brown: nucleoli.
Tan to brown: cytoplasm

Munoz, E., and I. B. Elfenbein (1973) *Arch. Path.* 96:111.

Silver for Mucopolysaccharides

A. PREPARATION

1.	Toluidine blue	1%
	Sodium carbonate	0.5%
2.	Boric acid (0.2 M)	50 ml
	Sodium borate (0.05 M)	59 ml
	Distilled water	91 ml
3.	Methenamine (3%)	8 ml
	Solution B	1 ml
	Silver nitrate (5%)	1 ml

(Solution 3 should be kept for ~10 days at 4°C in the dark.)

4.	Eosin	10 g
	Erythrosin	1 g
	Alcohol (25%)	100 ml

B. METHOD

1. Stain sections collected on a glass slide with a few drops of Solution 1 by heating on a hot plate (80°C) for 1–2 min.
2. Rinse thoroughly in distilled water and counterstain with Solution 3 at 80°C for 1–3 min.
3. Rinse thoroughly in distilled water and counterstain with Solution 4 at 80°C for 5–10 sec.
4. Rinse in distilled water, dry, and mount a coverslip.

C. RESULTS

Brilliant orange: mucopolysaccharides.
Violet-blue: nuclei.
Pale violet: cytoplasm.

Biagini, G., P. Borsetti, and R. Laschi (1972) *J. Submicro. Cytol.* 4:283.

Thionin/Acridine Orange

A. PREPARATION

1.	Thionin	0.5 g
	Distilled water	200 ml
	0.1 N NaOH	50 ml
	90% ethanol	250 ml
2.	Acridine orange	2.5 g
	Distilled water	200 ml
	0.1 N NaOH	50 ml

B. METHOD

1. Cut 0.3–2.0-μm-thick sections and stain for 5 min. in Solution 1 at 70°C. Sections thinner than 1 μm require longer durations.
2. Rinse thoroughly in distilled water and counterstain for 1 min. in Solution 2 at 70°C.
3. Wash in distilled water, dry at 70°C on a hot plate, and coverslip with a synthetic resin.

C. RESULTS

Blue to black: nuclei.
Dark blue: myelin sheath and granules of mast cells.
Yellow: collagenous and elastic components, secretion of goblet cells.

Sievers, J. (1971) *Stain Technol.* 46:195.

Toluidine Blue

A. PREPARATION

1. Stock solutions
   1% toluidine blue, aqueous
   25% sodium bicarbonate, aqueous (Na_2CO_3)
2. Stain

1% toluidine blue	1 ml
2.5% Na_2CO_3	20 ml

Prepare fresh as needed and filter before use.

B. METHOD

1. A drop of freshly prepared stain is placed over the epoxy sections on a glass slide. Allow to stain 0.5–3 min at 47°C using a microscope to check the degree of staining.
2. Rinse with distilled water. Slight destaining can be obtained with ethanol.
3. Dry and mount. Polymerization of epoxy resins by heat will cause some destaining.

Toluidine Blue/Acid Fuchsin

A. METHOD

1. Cut 1-μm-thick sections and heat-fix them on a clean glass slide.
2. Stain with 0.5% toluidine blue in phosphate buffer (pH 7.0–7.4) for 5–8 min. at 85°C.
3. Wash in tap water and oxidize with 2% periodic acid for 0.5–2 min. at room temperature.
4. Wash and stain with 0.5–1.0% acid fuchsin for 5–8 min. at 85°C.
5. Wash, dry, and mount.

B. RESULTS

Dark blue: nuclei.
Bright blue: muscle.
Red: collagen.
Deep red: elastica.

Removal of Reduced Osmium

I. Oxone

Oxidize sections in 5% aqueous Oxone for 1–4 hr at room temperature.

Sevier, A. C., and B. L. Munger (1968) *Anat. Rec.* 162:43.

II. Potassium Permanganate–Oxalic Acid

Expose the sections to 0.1% aqueous $KMnO_4$ for 1–3 min.; then bleach with 1% aqueous oxalic acid.

Shires, T. K., M. Johnson, and K. M. Richter (1969) *Stain Technol.* 44:21.

III. Performic Acid

A. Oxidizer solution

Hydrogen peroxide (30%)	5 ml
H_2SO_4 (conc.)	0.5 ml
Formic acid (97%)	44.5 ml

B. METHOD

1. Oxidize for ~10 min. with freshly prepared solution.
2. Stain with alcian blue–PAS–orange G or aldehydefuchsin–orange G.

Heath E. (1970) *Z. Zellforsch.* 107:1.

IV. Hydrogen Peroxide

A. Oxidizer solution

Distilled water	30 ml
H_2O_2	15 ml
0.01 N H_2SO_4	1.8 ml

B. METHOD
Oxidize for ~1 min. at room temperature with fresh solution.

Resin Removal Methods

I. Sodium Hydroxide: Epoxy Resins

1. Prepare a saturated solution of sodium hydroxide in ethanol and age it 3 days.

2. When the solution turns brown, immerse slides with sections and agitate 1–3 min.
3. Rinse in absolute ethanol.

Lane, B. P., and K. L. Europa (1965) *J. Histochem. Cytochem.* 13:579.

II. Sodium Methoxide: Araldite

A. SOLUTION (prepared in a fume hood)

1. Add small pieces of 2.5 g metallic sodium to 25 ml methanol at 50–60°C. Maintain the level of methanol at 25 ml.
2. After dissolution, add 25 ml benzene.
3. Store in an amber bottle.

B. METHOD

1. Expose the slide for 1–3 min. to the stock solution.
2. Rinse in 1:1 benzene and methanol.
3. Rinse twice in acetone.
4. Rinse twice in distilled water.

Mayor, H. D., J. C. Hampton, and B. Rosario (1961) *J. Biophys. Biochem.* 9:909.

III. Potassium Hydroxide: Epon

1. Prepare a saturated solution of potassium hydroxide in absolute ethanol.
2. The yellow-brown supernatant is the solvent.
3. Immerse slide 10–15 min.
4. Rinse in four changes of absolute ethanol.

Imai, J., A. Sue, and A. Yamaguchi (1968) *J. Electron Micros.* 17:84.

IV. Halogenation with Bromine

1. *Caution*: Bromine is extremely caustic and must be handled with care.
2. Expose sections to Br vapors ~30–60 sec.

3. Rinse in several changes of ethanol.
4. Rinse in distilled water.

SUPPORT FILMS

Formvar or Collodion Cast on Glass

1. Prepare a stock solution of 0.25% Formvar in ethylene dichloride or 0.75% Collodion in ethyl acetate.
2. Clean a microscope slide, leaving a slight detergent residue. Air-dry.
3. Dip the slide into the plastic solution and air-dry at an angle.
4. Fill an evaporation dish or fingerbowl to overflowing with distilled water, and roll a Teflon or glass rod over the water surface.
5. Scrape the edges of the slide with forceps, or score ~2 mm in from the edge with a razor.
6. Prop the slide at an angle (~30°) at the lip of the evaporation dish, and slowly push the slide into the water, leaving the plastic film to float on the surface.
7. Slightly bend clean 300-mesh Cu grids and place them (matte side down) on colorless, undamaged areas of the film.
8. Position a wide-mouthed test tube over the floating film, push it gently into the water, rotate the tube over the floating film, and remove it from the dish.
9. Air-dry the test tubes with attached grids under cover. Remove grids when dry.

Schaefer, V. J., and D. Hasker (1942) *J. Appl. Phys.* 13:427.

Collodion Cast on Water

1. Prepare a stock solution of 0.5–1% Collodion in ethyl acetate.
2. Position a clean wire mesh inside the base of a flat-bottomed funnel and fill the funnel with distilled water.
3. Place clean 300-mesh Cu grids (matte side up) on the wire screen.
4. Place a drop of the stock solution in the center of the dish above the water level, wait for the solvent to evaporate, and discard the film.
5. Repeat step 4, but do not discard: this is the support film.

6. Slowly open the funnel stopcock permitting water to drip out of the funnel.
7. The film will layer over the grids. Dry them, and remove them individually.

Ruska, H. (1939) *Naturwissenschaften* 27:287.

Carbon Support Films: Mica

1. Cut sheet mica into 1" × 1" squares; then cleave the sheet using a sharp razor blade. Each freshly exposed surface may be used as a target; these should be prepared just prior to use.
2. Preparation of bell jar:
   a. Position two carbon electrodes in the brass holders, making sure that the pointed electrode is spring-held and there is tension between it and the squared-off electrode.
   b. Clean a white porcelain dish, place a drop of vacuum oil on it, and place this to the side of and beneath the electrodes.
   c. Put the freshly cleaved mica into the bell jar 10–15 cm beneath the electrodes.
   d. Evacuate to $\geqslant 10^{-4}$ Torr.
3. Evaporate with an alternating current of ~30 amp at 15 V until the porcelain surrounding the drop is light brown; the film thickness is ~70 Å. Return to atmospheric pressure.
4. Using forceps remove the mica from the bell jar, and gently lower it onto the surface of distilled water in an evaporation dish; the angle should be $\leqslant 15°$.
5. Lower clean 200- or 300-mesh Cu grids individually beneath the water surface, position them beneath the film, and pull them up and through it.
6. Blot the grids, surface side up, on filter paper.
7. Apply the specimens.

NEGATIVE STAINING

Preparation of Negative Stains

1. Phosphotungstic acid
   1–5% at pH 6.8–7.4; 1% aqueous at pH 6.5 is standard. Titrate to desired pH using 0.25 M KOH.

2. Uranyl oxalate
   A solution containing 0.5% (12 mM) of oxalic acid is adjusted to pH 6.5–6.8 with dilute NH_4OH. 0.5–1% at pH 4.5–7.
3. Ammonium molybdate
   1–3% at pH range 6.0–8.0; 1% solution at pH 6.5 is standard.
4. Sodium silicotungstate
   1–4% at pH 6.8–7.2.
5. Uranyl acetate
   0.5–1% at pH 4.5–5.2; 0.5% at pH 4.0 is standard.
6. Uranyl formate
   0.5–1% at pH 4.5–5.2. Titrate to desired pH with NH_4OH.

Negative Staining: Drop Method

The standard procedure for staining above the isoelectric point with phosphotungstic acid, ammonium molybdate, uranyl oxalate, and sodium silicotungstic acid is as follows:

1. Grids mounted with plastic or carbon supports are picked up by the edge with forceps and placed on a flat surface with the support side up.
2. The specimen is diluted with water or low-concentration buffer to the appropriate concentration. The ideal concentration depends upon the type of specimen, stain and concentration, and the type of support film.
3. A small droplet of the sample is placed on the support to form a bead nearly extending to the edge of the grid. A fine-tipped micropipette or platinum loop can be used.
4. The droplet is touched with the torn edge of ashless filter paper to remove most of the excess liquid, and a drop of stain is immediately applied before the residual film of liquid can dry.
5. After a discrete length of time (30 sec is standard), remove the stain by touching in with torn filter paper.

Negative Staining: Float Method

1. Follow steps 1 and 2 as above in the "Drop Method."
2. Place a drop of the sample suspension in a plastic petri dish; or,

position a piece of parafilm in a petri dish, and place a drop of the suspension on the parafilm.

3. Float the grid, support side down, on the drop for ~30 sec. Low concentration suspensions require longer durations.
4. Remove the grid, blot off excess sample, and immediately negatively stain the sample by floating on a drop of stain.

Author Index

Subject Index